Charles Blacker Vignoles:
romantic engineer

Charles Blacker Vignoles: romantic engineer

K.H. VIGNOLES

CAMBRIDGE UNIVERSITY PRESS

CAMBRIDGE

LONDON   NEW YORK   NEW ROCHELLE

MELBOURNE   SYDNEY

Published by the Press Syndicate of the University of Cambridge
The Pitt Building, Trumpington Street, Cambridge CB2 1RP
32 East 57th Street, New York, NY 10022, USA
296 Beaconsfield Parade, Middle Park, Melbourne 3206, Australia

First published 1982

Printed in Great Britain at the University Press, Cambridge

Library of Congress catalogue card number, 81–10226

British Library cataloguing in publication data
Vignoles, K.H.
   Charles Blacker Vignoles.
   1. Vignoles, Charles Blacker   2. Civil
   engineers – Great Britain – Biography
   I. Title
   624′. 092′4   TA140.V/
   ISBN 0 521 23930 3

CONTENTS

NOV 3 1982

ILLUSTRATIONS

Illustrations

TO THE MEMORY OF
ARTHUR ELTON

PREFACE

The suggestion that I might write this book was made some years ago by the late Sir Arthur Elton, whose untimely death early in 1973 was so great a loss to his friends and to lovers of railway history. He was doing some research into the connection between Vignoles and the artist John Cooke Bourne, and he wrote to ask whether I knew the whereabouts of Vignoles's diaries. At the same time he suggested that there was a need for a new biography of Vignoles, and enquired whether perhaps I was writing one.

Some years before, the Vignoles family correspondence had come into my possession, and I had learned of the existence of Vignoles's extensive diaries in the Manuscript Department of the British Library. The opportunity to study these papers, afforded by retirement, had stimulated my interest in my great-grandfather's life story, and encouraged me to think seriously of Arthur Elton's suggestion, particularly as he offered to help me to find a publisher. The whole affair was clinched a few months later when he spent a night at my house inspecting Vignoles's papers. Neither my wife nor I will ever forget the occasion. From the moment when he descended from the train at Emsworth station, a burly bearded figure clutching a bulging carpet bag, to the time he left us next morning, except for the intervals necessary for sleep and meals, he was fully occupied, closely examining the stream of letters and papers we put before him, and exclaiming with boyish delight at everything he saw. It was an example of scholarly enthusiasm so infectious that we were spellbound, and the memory still makes it hard to believe in the fact of his death. Among the many items of help and advice he gave me were the opportunity to examine his collection of railway prints and his library, during a weekend my wife and I spent with him and Lady Elton at Clevedon, and the loan for several months of a precious copy of the report of the Irish Railway Commission. For all this I am deeply indebted to him, as well as for the many fascinating hours I have spent on this project. My only regret must be that he did not live to see the work completed. He shared with Vignoles the same quality of unbounded enthusiasm; and I am proud and pleased to be able, with Lady Elton's permission, to dedicate this book to his memory.

Owing to pressures arising from inflation, the original publishers were unable to complete the work, and for some years the book lay fallow, until it was accepted by the Cambridge University Press, to whose staff I am indebted for advice on a considerable revision of the text.

Preface

Vignoles's letters and diaries are rich material for the biographer. His son, the Rev. Olinthus J. Vignoles, quoted extensively from them in his *Life of Charles Blacker Vignoles*, published in 1889, hitherto the main authority on Vignoles for historians and students of railway history. He declared his intention to allow his subject 'wherever possible, to tell his own story'. But absence of relevant comment and explanation often leaves the story obscure; and as Olinthus Vignoles does not hesitate to alter the text of his quotations when he thinks fit, the character of his father's vigorous personal style is often lost and the facts distorted. In other respects, however, the portrait he draws is reasonably balanced, and his book contains some fascinating reminiscences of early railway history.

Starting with the framework provided by O. J. Vignoles, I have based my story of Vignoles's life on a complete re-examination of the letters and diaries, backed up wherever possible by reference to contemporary records, such as newspaper reports and the minutes of railway companies, and to the works listed in the bibliography. These have invariably confirmed the essential accuracy of Vignoles's own account, so that where the latter is the only available source of information (for example in the story of the building of the Kiev Bridge) I have been able to follow it with some confidence. My account of Vignoles's ancestry, which differs sharply from that given by O. J. Vignoles, is based on the patient research carried out over several years by my uncle, the late Ernest B. Vignoles, and other members of the family.

With so much material available, the task of selection has been a formidable one. The temptation to dwell on many fascinating details of personal and family history, in what is primarily intended as a study of an engineer's career, has had to be resisted; though I have thought it proper to retain sufficient of these to illuminate the personal background of a professional man in action. On the engineering side I must pay tribute to the invaluable assistance given me by my son, Mr M. J. P. Vignoles, M.A., C.Eng., M.I.C.E., who has acted as my technical adviser in all matters concerning Vignoles's professional work.

Passages in inverted commas, unless otherwise assigned, are taken from Vignoles's own writings, and in all cases the original spelling and punctuation has been retained.

I would like to thank the librarians, at home and overseas, who have been so ready to answer enquiries, and whose patience in ferreting out information seems quite inexhaustible. I am also grateful for the facilities for work and study so freely given by the libraries to anyone engaged in research. Foremost among these I would mention the Department of Manuscripts of the British Library, the Library of the Institution of Civil Engineers, and the British Transport Historical Records Department of the Public Records Office. I have also to acknowledge the help given to me by members of the staff of the following: the Royal Astronomical Society; the Reading Room and the Newspaper Library of the British Library; the Wills Memorial Library of Bristol University; the Cheltenham Public Library; the Hampshire County Library; the Lancaster District

Central Library; the Library of the Institution of Mechanical Engineers; the Library of the Royal Military Academy, Sandhurst; the Central Library and the Literary & Philosophical Society, Newcastle-upon-Tyne; the Royal Photographical Society of Great Britain; the Portsmouth City Library; the Portsmouth City Record Office; the Harris Museum & Art Gallery, Preston; the Wardens of Rochester Bridge; the St Helens Central Library; the Library of the Science Museum, London; the National Library of Ireland; the Public Record Office of Ireland; the Records Department of Coras Iompair Eireann, Dublin; the New York Public Library, Prints Division; the Charleston County Library, South Carolina; the St Augustine Historical Society, Florida; the Technical Library, Leningrad. I am also grateful to the following individuals for advice, encouragement or help: Mr Anthony Adams, of the Moonraker Press, Bradford-on-Avon; Mr J. S. Allen, of the John Thompson Horseley Bridge Co. Ltd; Mr Edgar Anstey; M. André Bodin; Dr L. G. Booth, of the Department of Civil Engineering, Imperial College of Science & Technology; Mr John Burnett; Professor W. H. Chaloner, Professor of Economic History, University of Manchester; Mr J. H. Colyer-Fergusson, of the National Railway Museum, York; Lady Elton; Mr J. C. Gilbert; Mr A. C. Gutteridge; Señor Adolfo Lafarga, of the Biblioteca Provincial, Bilbao; Mr J. J. Leckey; the late Mr Peter Mrosovsky; Mrs Mary Monro, formerly of Thomas Telford (Publishing) Ltd; Mr K. A. Murray, of the Irish Railway Record Society; Mr R. H. Offord, of Manchester University Press; M. André Portefaix, of the Revue Générale des Chemins de Fer, Paris; Mr N. Rayman, City Engineer, Coventry; the late Mr L. T. C. Rolt and Mrs S. M. Rolt; Professor J. Simmons, of the Department of History, University of Leicester; Mr Jack Smythe; Mr C. W. Toogood, of the Oxford University Press; Señor Alfonso C. Saiz Valdivielso, of the Bank of Bilbao; Miss Margaret Wallace; and Professor J. M. Wallace-Hadrill.

I am indebted to the Editor of the Journal of the Irish Railway Record Society, for permission to quote from the transcript of Richard Osborne's diary; acknowledgements are also due to the Editors of the *Railway World* and *Country Life*, in whose pages some sections of Chapters 3 and 5 respectively have already been published.

Permission to publish, given by the various bodies mentioned in the list of illustrations, and to quote from unpublished material, is also gratefully acknowledged.

Finally I would like to thank Dr Simon Mitton of the Cambridge University Press, for his great interest, courtesy and encouragement; I remember with gratitude the hospitality afforded to my wife and myself by my cousin the late Mrs Lydia Vignoles de Ward, during our visits to London for research; and I thank my wife for her help and companionship in research, for her encouraging and constructive criticism, and for typing the final draft of the book.

K.H.V.

VIGNOLES AND HUTTON FAMILIES

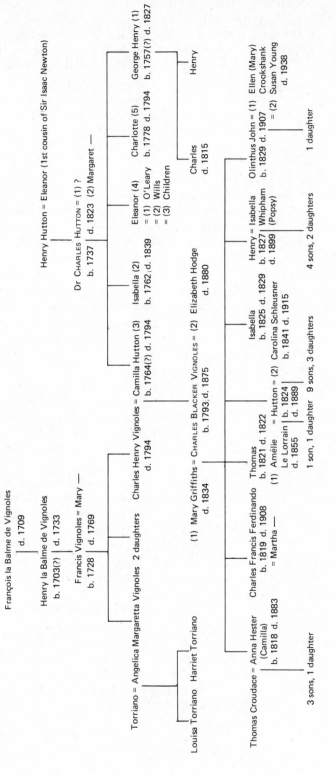

1 The infant ensign, 1793–1817

In late summer 1795, three oddly assorted travellers arrived at the house of Dr Charles Hutton, F.R.S., Professor of Mathematics at the Royal Military Academy, Woolwich. One of them was the Doctor's son, Henry, a captain in the Royal Artillery; the others, Captain Hutton's two-year-old nephew, and an Irish nursemaid. Together they had made the long journey across the Atlantic from the West Indian island of Guadeloupe.

This was the climax of a series of events which had begun five years before, when Dr Hutton's second daughter, Camilla, had married Captain Charles Henry Vignoles, of the 43rd Regiment of Foot. Vignoles was a descendant of a line of Huguenot soldiers serving in the British Army.[1] Shortly after the marriage, he joined his regiment in Ireland, and his wife went with him. During the various changes of lodging necessitated by the regiment's movements, the couple were invited to stay for a time by some friends of her husband, a Mr and Mrs Blacker, at their country house at Woodbrook, near Enniscorthy, County Wexford; and it was in the Blackers' house, on 31 May 1793, that Camilla gave birth to a son, who was christened Charles Blacker out of compliment to his parents' host and hostess.

The child was barely six months old when his father's regiment was ordered overseas. Britain's struggle with revolutionary France was just beginning, and a combined operation was to be launched against the French West Indian islands, from a base in Barbados. The 43rd embarked at Cork on 17 November and, as was by no means unusual in the eighteenth century, Captain Vignoles was accompanied on the seven weeks' voyage to the Caribbean by his wife and son. What such a journey must have involved, for a young mother nursing her firstborn child, is hard to imagine.

Presumably Camilla found lodgings in Barbados, and remained there while the campaign against the French islands proceeded. At first all went well. Martinique and St Louis were taken, not without heavy fighting, and some time after the fall of Guadeloupe, in April 1794, the Vignoles family were reunited in the town of Pointe-à-Pitre, capital of the island. But here disaster struck them. Yellow fever, the scourge of all such campaigns, had broken out among the troops and seamen. The British forces, depleted even more by sickness than by their battle casualties, were unable to resist an unexpected attack by a force newly arrived from France. Pointe-à-Pitre was captured by the French revolutionary troops and Captain Vignoles, wounded and suffering from yellow fever, took refuge with his wife and son in the house of a friendly French merchant called Courtois.

In spite of all that M. Courtois and Camilla could do for him he died on 8 June. His wife, also suffering from the fever, and exhausted by the strain of nursing her husband, died two days later. In a letter to Dr Hutton, M. Courtois tells how Camilla, who wished to write to her father and brother before she died, 'asked for ink and paper, but her strength failed, and she could do no more than leave us their address . . . she ended by confiding to my care her child Charles, then thirteen months old, who was already so ill that he seemed unlikely to survive her long. I promised to take care of him until the coming of peace would allow me to return him to his family.'

M. Courtois was as good as his word. The baby was nursed back to health, and a message was somehow conveyed to Camilla's brother Henry, then serving with the artillery in Grenada. Having obtained per-

Captain Charles Henry Vignoles, *c.* 1790, from a miniature by an unknown artist.

mission to go to Guadeloupe in search of his nephew (his intention was to pass through the French lines under a flag of truce), he joined the remnants of the British forces holding out in the island. An action ensued, in which Captain Hutton was wounded in the eye and taken as a prisoner to Pointe-à-Pitre, where he was not only able to locate his nephew, but managed to persuade the French commander to release him on parole, so that he could return to Grenada with the child and his nursemaid, and thence to England.[2] Meanwhile, in response to an application from Captain Hutton, the British Commander-in-Chief in the West Indies had appointed the infant Vignoles to the rank of Ensign in his late father's regiment, on condition that 'he shall exchange to half-pay immediately, being too young to serve'.

The suggestion in the *Dictionary of National Biography* that the commission was bestowed on Charles in order that he might be exchanged for a French prisoner of war, is not supported by a close examination of the evidence. Olinthus Vignoles, on whose biography of his father the Dictionary largely draws, while mentioning the condition of Captain Hutton's exchange, merely points out that the grant of a half-pay commission to an infant was a means by which the War Office sometimes compensated a deceased officer's family. Romantic as the Dictionary's suggestion is, it must be discounted.

The family to which the infant ensign had been so miraculously conveyed consisted of Dr Hutton, Margaret, his second wife, and his eldest

Isabella and Camilla, daughters of Dr Charles Hutton, 1788. Oil painting by Philip Reinagle.

daughter Isabella, then in her thirty-third year. The son, Henry, we have already met. A third daughter had already married and left home, while another had died the previous year at the age of 16, about the same time as the tragic news from Guadeloupe reached Woolwich. It had been reported that young Charles had perished with his parents;[3] so that his unexpected restoration must have seemed to the Huttons like a heaven-sent consolation for the loss of his mother and aunt. Small wonder that the two ladies took him to their hearts. As for the Doctor, there is no doubt that Captain Hutton was carrying out his sister's wishes in delivering the child to him, and he readily accepted full responsibility for his upbringing.

Dr Charles Hutton, at 58 years of age, was at the height of his powers as Professor of Mathematics at the Royal Military Academy. For 20 years he had been a Fellow of the Royal Society; he was an authority on ballistics, gunnery and the theory of explosives; he had calculated the mean density of the earth from the data provided by Maskelyne's experiments on Mount Schiehallion, and he was the author of numerous works on

Dr Charles Hutton F.R.S., aged 85 years, in his house at 34 Bedford Row, 1822. Oil painting attributed to Philip Reinagle.

mathematics. The son of a Newcastle colliery official, he had worked as a boy in the mine; but when an accident made him unfit to hew coal he was sent to the village school, where his considerable mathematical ability brought him to a career of teaching. At 23 he had founded his own mathematical school at Newcastle, and only 13 years later he was appointed Professor at Woolwich. Nor was this all. In 1772 he had published a book on the principles of bridges, after the Newcastle bridge had collapsed in a flood; and he had been commissioned by the city authorities to survey and map the city and its suburbs. At Woolwich he had designed and built 18 houses on a piece of ground he had bought on the wind-swept slopes of Shooter's Hill. In one of these houses he made his home.[4]

Apart from a couple of brief references in family correspondence, the little we know of Charles's childhood is based on some passages in letters he wrote in 1814 to his future wife. These tell us that from an early age he enjoyed the advantages of his grandfather's skill as a teacher. He was well grounded in mathematics, classics and modern languages, and the Doctor gave him the run of his well-stocked library. Here Charles read widely and voraciously, preferring literature to books on science, and on his own admission he spent more time browsing and dreaming in the library than at more methodical studies. Such an attitude could hardly have been approved by his grandfather, who was noted for his undeviating regularity in the distribution of his time, and, though generally of a mild and equable temper, was a man of authority in every sense of the word. But it is probable that Charles found sympathetic support from the ladies, and that they encouraged his undoubted artistic talents, and showed understanding for a clever boy's natural aversion to drudgery.

From his reading Charles confessed that he remembered the ornamental rather than the useful; but his close acquaintance with the classics, Shakespeare and the literature of the eighteenth century, and his knowledge of languages, taught him to write in a style which, though not uninfluenced by the fashionable verbosity of the age, yet displayed wit, imagination and a feeling for good English. Having the example of his grandfather's skill as a draftsman, his forte was in mechanical rather than freehand drawing, though he seems to have shown promise in sketching and painting. Add to all this a well-trained singing voice and a practical knowledge and appreciation of music, and the picture emerges of a cultivated and intelligent young man, typical of his age, whose upbringing had bred in him a considerable independence of spirit.

Before Charles completed his education, Dr Hutton retired from the Academy, and the family moved to London. The new home was at number 34 Bedford Row, a quiet street whose pleasant eighteenth-century houses still face one another within a stone's throw of Gray's Inn. The move to Bloomsbury put the Doctor within easy reach of his wide circle of scientific and literary acquaintances, and of many army officers (including some of the highest rank) whose friendship he enjoyed through his long association with the Academy.

For the youthful Charles, halfway between boy and man, the change meant a considerable widening of horizons. It was not long before he was taking part in social evenings at Bedford Row and in the other Bloomsbury houses of his grandfather's friends. Poetry readings and music, lectures and discussions on scientific subjects, were the main features of these gatherings; and it was here that he first began to be aware of the charms of the opposite sex, with whom he sharpened his wits in argument, and tried his hand at writing acrostics and love verses.

As Charles grew to manhood he continued to work at mathematics with his grandfather, assisting him with the computation of a set of logarithmic tables. Although his grandson was still an ensign on half-pay, and the army was deeply involved in a continental war, the Doctor had no intention of allowing Charles to take up a full commission. Instead, he proposed he should enter the law. Charles was accordingly articled to a proctor in Doctors' Commons. He seems to have accepted this, though how willingly we do not know. It is possible that Dr Hutton regretted having allowed his daughter to marry a penniless officer, and that he wished to make amends for this, and for her untimely death, by ensuring a more secure life for her son. Peace must come ere long, and after 30 years at the Academy he had no illusions about the prospects of peacetime soldiering without influential support and considerable private means. So he probably decided that such money as he was prepared to bestow on his daughter's son would be better invested in a legal training than in an uncertain military career.

We do not know how long Charles stayed in Doctors' Commons, but the mental discipline of this legal training, and the knowledge of the law he acquired there, were to be of great value to him in his professional life. Meanwhile the delights of the social gatherings in Bedford Row were some compensation for the drudgery of a lawyer's office. To these Charles contributed his share of 'poetic effusions', of which two survive: *The Sylphiad*, a leather-bound manuscript poem in eight cantos, written in elegiac couplets, and owing more than its inspiration to Pope; and a shorter heroic ode, entered for a competition held in August 1812 for a poem to be read at the re-opening of Drury Lane Theatre, rebuilt after a disastrous fire. The ode is preserved in the British Library, bound up with the hundred or so other unsuccessful entries (with many of which it compared favourably) under the title of *Rejected Addresses*.[5] (The poem actually spoken at the re-opening was written by Byron, who had not even deigned to enter the competition.)

About the middle of 1813, when Charles was just 20 years old, this idyllic phase of his life came to an abrupt end. He quarrelled violently with his grandfather, and was banished from the house. What exactly happened is wrapped in mystery; but the result was a rift between grandfather and grandson that had an influence on the whole course of the latter's life. According to Charles's account, he was engaged in a clandestine love affair, and had for the first time in his life run into debt (it was by no means to be the last), 'when the thunderbolt burst over my head and I left home'. In later years Mrs Hutton told Charles that the Doctor could never forgive him for 'going from the law business'. It seems probable, therefore, that Charles was beginning to tire of the law, and when taxed with his misdemeanours was unwise enough to tell his grandfather so, and to declare he wished to follow the Vignoles profession of arms; and that the Doctor, furious at having his plans thwarted (and incidentally at having laid out a large sum of money to no purpose) broke with him irrevocably. A reason for the break can also be found in the fundamental incompatibility between the Hutton and Vignoles temperaments. North country prudence, industry and hard-headed economy accorded ill with the more volatile and easy-going ways of the Huguenot character. In Charles the two opposing strains had met, to become a source of strength as well as weakness. His success would depend on which of the strains had the upper hand. In his grandfather's eyes the improvident strain had already taken over; what he could not foresee was that the breach would be made the more difficult to heal because of the very traits inherited by Charles from himself.

Yet the Doctor did not cut Charles off altogether. At the end of September 1813 we find him living at Sandhurst as a private pupil of Professor Thomas Leybourn, a teacher of mathematics at the Royal Military College. It seemed that Dr Hutton had consented to his grandson's wishes to the extent of putting him in Mr Leybourn's charge, with the injunction that he should do what was necessary to prepare him for the army and procure him a commission. There is no record of his ever having been a pupil at the College, although he mixed with the cadets, and he mentions having been drilled by the Regimental Sergeant-Major. As the holder of a half-pay commission, of more than 18 years' seniority, he had no need to go through the formality of purchasing a commission. It was merely a question of obtaining the favour of a senior officer willing to accept him in his regiment. Dr Hutton hoped that this might be effected by an application to the Duke of Kent, on whose staff Charles's father had served during the West Indian campaign. Meanwhile Mr Leybourn would see that Charles was kept out of further trouble.

What happened between the break with Dr Hutton and Charles's establishment at Sandhurst? There is a gap in the record here which one would dearly like to fill. Olinthus Vignoles refers to a tradition that his father joined the army in Spain and was present at the battle of Vitoria, and he goes on so far as to state that in later life Charles often spoke of an early visit to Spain. The story was sufficiently well-known for it to be referred to in a speech made by a distinguished Member of Parliament, when proposing Vignoles's health at the 1870 annual dinner of the Institution of Civil Engineers; and it is mentioned in various forms in the many obituary notices published in December 1875, most of which fasten on the picturesque circumstance that the young campaigner who 'threaded the passes of the Pyrenees' after the battle of Vitoria used this experience in later life when building a railway in the same area. But it is a fact that Olinthus Vignoles wrote at least one of these notices, and probable that he supplied material for the rest, many of which present a very garbled version of other incidents in Charles's life.

What is certain is that while the veteran engineer does not seem to have contradicted the statement made at the Institution dinner (which in any case he could hardly have done without discourtesy and some loss of face), yet the young soldier, in letters written to his fiancée much nearer the time, made no reference, direct or oblique, to an experience so recent and of such an adventurous nature. Such reticence is scarcely in character. Nor is there any trace of the story in any other family letters.[6] Furthermore, if Charles had in fact been with Wellington's army in the Peninsula would he have told his fiancée, in writing on the assault of Bergen-op-Zoom in March 1814, that it was there that he was first inured to fire? And is it likely that a young man who had already seen active service overseas would have submitted to the restraints put upon him under the tutelage of Mr Leybourn? In spite of all this the tradition is a persistent one.

However uncertain we may be about this particular episode, with Charles's arrival at Sandhurst we enter one of the best-documented periods of his life. For it was under Mr Leybourn's roof that he met Mary Griffiths, to whom he was to be secretly engaged for four years before finally marrying her, and with whom he was to carry on a long and intimate correspondence. She was the eldest daughter of a Welsh gentleman-farmer, after whose death she had set up as a milliner in London, with her two younger sisters in her care. Mr Leybourn was her guardian and trustee.

Charles, by his own account, was still smarting at the humiliation of having been banished from home and put in Mr Leybourn's charge as an unreliable good-for-nothing. The sympathy of an attractive young woman was just what he needed. On Mary's side, the warning that her guardian seems to have given her about the handsome and gifted young man's dangerous character, far from discouraging her, aroused her interest. She was six years older than Charles, and her strong maternal instinct persuaded her that he was a brand to be plucked from the burning. In his susceptible state, Charles was ready to be saved, while Mary's innate prudence proved of no avail against the penitent's undoubted charm. By the time she had to return to London, at the end of September, they were secretly engaged to be married.

There was need for secrecy, for Dr Hutton would hardly have approved such a match for his grandson, and Mr Leybourn must perforce have been of the same opinion. In any case, without a reconciliation with his grandfather, Charles could not hope to enjoy the 'competence' necessary for a stable marriage; he had still to be established in the army, and even then, without some private means, he could hardly hope to maintain his wife in the state they both would have wished. The prospects for the marriage were bleak. The lovers were separated as soon as they were engaged, and their meetings for the next four years were to be few and far between. In the circumstances it was remarkable that Charles, who had such a reputation for unreliability, should have remained faithful for so long.

Their correspondence was carried on in an emotional and highly romantic style, spiced with the excitement and alarms arising from the need for secrecy. Charles and Mary were true children of their age. Pro-

testations of love, confessions, misunderstandings, quarrels and recon-
ciliations, all the ingredients of the romantic novel of the eighteenth century, are to be found in these letters. They present a remarkable picture
of the daily life and character of the writers. They also reveal only too
clearly the incompatibility of temperament which was ultimately to wreck
their marriage. Charles and Mary were not the first couple to find the
realities of married life somewhat different from the romantic dreams of
courtship, especially when extended over a long period of separation.

The promised interview with the Duke of Kent took place, and after
some weeks, during which Charles pined for Mary in the country around
Sandhurst, 'rambling through every walk where we strolled together, and
resting on every Style and Gate which supported us both', he was commissioned to the York Chasseurs, situated in the Isle of Wight. There,
after trudging for a month through heavy snow on outpost duty at San-
down, he obtained a transfer to the 1st Royals (the Royal Scots) and was
ordered to join the 4th Battalion in Holland, where a British force was

Letter from
Charles Vignoles
to Mary Griffiths,
13 May 1814,
written on board
H.M.S. *Leopard*,
off Cork. A good
example of a
'crossed-over'
letter.

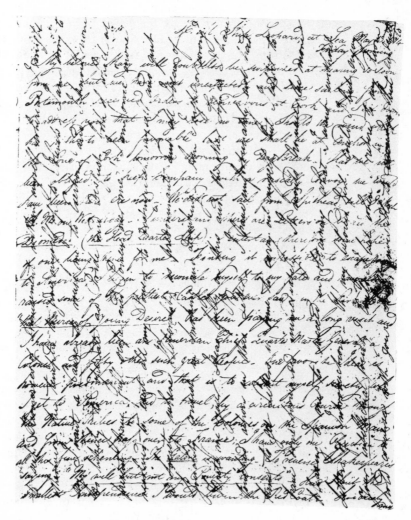

to reinforce a Dutch rebellion against Napoleon. At this point he must have done some hard thinking about his prospects as an infantry officer, for on arrival at the army base at Willemstadt, he applied for a post on the staff as assistant engineer. For this he confidently quoted as his qualification an 'engineering education', and a knowledge of French, German and grammatical Dutch, of which he knew enough 'to be sure he would soon be able to converse in it'. With equal confidence he tackled the job of making a plan of the town and fortress, which was required to support his application. 'Here was a task for me,' he wrote to Mary, 'who (between ourselves) knew nothing of affairs of that kind and was unfurnished with a single Instrument or Book to assist me. That, however, did not deter me.' In five days, with pencil, paper and his feet as measures, he had taken the dimensions of the ramparts and town and made a plan with the help of a pair of old compasses.

This achievement, and the information that he had learned his engineering from Dr Hutton, secured him the post; but when the appointment was confirmed in General Orders his commanding officer refused him permission to leave the regiment.

Next day, on the night of 8 March 1814, the Royals took part in the disastrous attack on the fortified town of Bergen-op-Zoom. Charles wrote to Mary a long account of the action, which included a vivid description of the abortive attempts to force the town after the troops had struggled across the River Zoom, waist deep in icy mud, and of the surrender of the British after a night of skirmishes and unsuccessful attacks. Charles had his cheek grazed by a musket-ball when leading a bayonet charge, and according to his account was the officer deputed to negotiate the terms of surrender for the battalion, carrying a white handkerchief (his own) tied on top of a halbert.

It was in this action, as we have already mentioned, that he told Mary that he was first 'broken in to stand fire'. The events of that dreadful night made a deep impression on him. He describes the guard-house in which the wounded found temporary shelter during the action:

The fire shone feebly on the ghastly countenances of the Men nearest to it. Their Cheeks streaked with Blood, the Eyes convulsed and every feature around distorted with Pain and Agony. In the background the Figures were scarcely discernible through the Smoke. The whole formed a scene that made my Heart sicken within me. No Pen could describe it, no Pencil do it justice.[7]

The terms of surrender provided for the regiment's return to England as soon as the ice-bound river was navigable, on condition that officers and men should not fight again against the French until exchanged for an equivalent number of French prisoners. While waiting to embark, Charles temporarily secured his staff job as assistant engineer, in the drawing department of the army headquarters at Breyda. Here he was 'living like a prince, with three horses and two servants in a magnificent billet – easy in mind and fat in body'. He planned to send Mary a consignment of lace and linen packed in a cask as 'damaged ammunition'.

On 4 April, with Napoleon's abdication, the war in Europe was virtually over; but Mary's relief at the news was short-lived; no sooner were the Royals back in England than they were ordered to embark for North America, where a British army based on Quebec was carrying on the campaign against the United States. Charles managed to snatch a couple of days with Mary in London, before sailing with the regiment from Portsmouth in H.M.S. *Leopard*, early in May 1814. Though still unreconciled with his grandfather, he was reminded in a letter from Mrs Hutton of the advantages a grandson of so important a person would always enjoy: 'Go to what quarter of the globe you may, you will always find someone acquainted with your grandfather, who for his sake will be kind to you.'

It was a long, slow voyage, in a leaky ship, but luckily Charles was a good sailor, and interested in seamanship and navigation. He wrote long letters to Mary, in one of which he speculated on his future, with a remarkable understanding of his own character:

Finding it impossible to escape going to America, I begin to reconcile myself to my Fate and have already raised some of the prettiest Castles you ever saw – in the Air! Hope, 'that Nurse of young Desire' has been paying me a long visit, and I have already been an American Chief, Quarter Master General, a Colonel and fifty other such great People. One favorite Idea . . . is to absent myself secretly when I get to America, and to travel, by a circuitous Course through the Native Tribes to some of the Colonies on the Spanish Main and from thence pass over to France. I have only one objection to all this fine scheming – the *Galore* is wanting. To reverse Shakespeare's saying 'My Will but not my Poverty consents'. Had I but the smallest Independence I would burn the Red Coat. Seriously, my sweet Mary, do you not think we could live happily in some sequestered Cottage in the Alps or the Appenines . . . In imagination the Scheme is heavenly, but I fear dearest Love it would soon lose all its charm. I am consciously too fond of a dash, a show-off, and I seriously begin to think it is the hope of one day being able to make one, that restrains me giving way to my present Inclinations: much more than a fear of offending my best friends: my Pride and not my Virtue, is the agent.

He celebrated his twenty-first birthday in a gale, and a month later the voyage came to an unexpected end, when the *Leopard*, feeling her way into the mouth of the St Lawrence in thick fog, struck a reef of rocks on the eastern end of the island of Anticosti. Although badly holed, she remained fast on the reef. The whole ship's company was landed successfully on this barren island, where sailors and soldiers had to bivouac in acute discomfort. Charles was one of two officers despatched with the sailing-master in a local fishing schooner to seek help from Quebec, where he arrived 17 days later, sick with severe inflammation of the lungs. It was not until 4 August that he was able to write to Mary, being still very weak and 'reduced to a skeleton'.

Charles's eight months' stay in Canada was militarily uneventful. He was not required to go up to the Front on the Great Lakes, possibly because of his health. His commanding officer, Colonel Muller, wrote to the Duke of Kent commending his behaviour in the shipwreck, but refused again to allow him to take a staff post as engineer. Communications with England were slow, a reply to a letter taking anything up to three months. Though he heard again from Mrs Hutton, who gave her opinion that the loss of the *Leopard* must have been due to 'some want of either skill or attention', no letter arrived from Mary until December. Charles wrote to her that he had grown reserved and retired and was spending the time he could spare from regimental duties in study.

But life at the garrison had its lighter moments too. Amateur theatricals were the regiment's chief diversion, and in the absence of talent or inclination among the ladies, Charles, who was below average height, made quite a hit as Julia in *The Rivals*. And the Duke of Kent's birthday was celebrated by a ball, for which Charles designed and painted a large transparent portrait of the Duke. He worked off his pique at Mary's long silence by describing the evening's delights in detail:

We kept threading the Rounds of the Mazy Dance till 5 o'clock in the Morning; at which time the Ladies having retired; and the Claret operating rather strangely, I with a few more gallanted the Housemaids &c. of the Hotel into the Ball Room and prevailing on the drowsy Musicians to strike up, finished the Night's Amusements with a few Reels; the last of which was a Reel to Bed; it could be called nothing else!

With the American war ended, and Napoleon finally defeated at Waterloo, the Royals were back in England. Charles was now posted to the 1st Battalion in Scotland, first in Edinburgh, and then to the remote highland outpost of Fort William, where he was put in command of the detachment on garrison duty. Here he was made welcome by the local gentry, and at Christmas was much in demand at balls and dinners for miles around. But though now a full lieutenant, the shadow of being relegated to half-pay hung over him, with the inevitable reductions brought about by the coming of peace.

The blow fell in April 1816; however he partially evaded it by securing a post as A.D.C. to General Sir Thomas Brisbane, with the army of occupation at Valenciennes. The future governor of New South Wales, who had wide scientific interests, gave his new A.D.C. scope for the exercise of his mathematical ability and skill as a draftsman; and when the Duke of Wellington demanded a comparative table of French and English weights and measures (including the new-fangled decimal system), it fell to Lieutenant Vignoles to produce it. But the post carried no pay and life at divisional headquarters was expensive. In January the announcement of wholesale reductions, involving the relegation of 1000 lieutenants to half-pay, plunged him into despair. Mrs Hutton sent him £5, but warned him against returning to England, where the country was in a state of unrest arising from the poverty and distress caused by the post-war slump. This was the last letter Mrs Hutton wrote to Charles before she died. He had never lost her affection, and she had kept the door, which his grandfather had slammed in his face, unlocked if not open.

But if England had nothing to offer, could not an ambitious young man find employment, and even fortune, overseas? Such an opportunity seemed to offer in the Spanish dominions of Central America, where British and American guerilla forces were supporting the cause of revolution, led by Simon Bolívar. Agents were at work in England, raising officers and men from the many thousands laid off from the army. Charles noted that engineers and artillery officers were particularly in demand. In May 1817 he returned to England determined to seize this opportunity, and he flung himself into the idea with typical self-confidence and enthusiasm.

Still faithful to Mary, whom he would have married on his return from Canada but for her continued reluctance to embark on a penniless marriage, he now urged her to marry him and go with him to South America. His romantic imagination envisaged golden prospects of a short war and the riches of Peru as a reward. But Mary hesitated; her practical nature saw the scheme as altogether too rash and dangerous. Charles spent several weeks in anguished uncertainty, at the end of which he had to report at Portsmouth to join a vessel due to sail early in July. He now discovered that women would not be allowed to travel in any case. At last Mary consented to marry him before he sailed, even though she could not go with him. Charles had taken rooms in Gosport while awaiting embarkation. On the night of 12 July Mary travelled down by the mail coach to Portsmouth, and early on Sunday morning – the 13th – they were married in Alverstoke Parish Church.

2 South Carolina and Florida, 1817–1823

A couple of days before their wedding, Charles wrote Mary a letter of precise and meticulous instructions for joining him. The timing of the night journey by mail coach, the price of the successive fare stages, the amount of the necessary tips to servants, the promise of a warm bath for her at the Crown Inn at Portsmouth to refresh her after the journey, nothing was forgotten which could give her comfort and reassurance in the momentous step she was taking. But the precision of these plans was in sharp contrast to the unstable foundations he was laying for their married life. He was leaving Mary for an indefinite period, himself bound on an expedition of doubtful outcome. Though Mary could enjoy some support from his half-pay, he had still many unpaid debts; yet pride prevented him from pleading for help from his grandfather, as well as the belief that once the Doctor discovered that he had married without his consent all hope of reconciliation would vanish. He agreed with Mary that prudence demanded that their marriage should be kept secret. On one thing how-

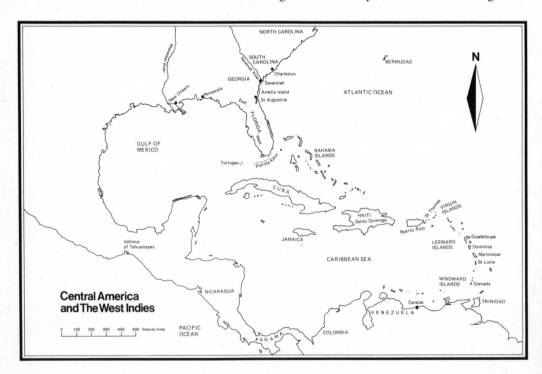

Central America and The West Indies

ever he insisted. She must join him in America as soon as it became safe for her to do so. This strange mixture of precise attention to immediate detail and hopeful confidence in an uncertain future was only too typical of Charles's character.

After a brief honeymoon in Gosport they parted. Mary to return sadly to London by coach, and Charles to embark on the *Two Friends*, in company with an ill-assorted company of individuals brought together by the promise of active and lucrative employment in the rebel forces in Central America. Some, like Charles, were navy or army officers on half-pay; the rest were less respectable, among them officers discharged after a court martial, failed midshipmen, run-away apprentices, and a fair sprinkling of good-for-nothing adventurers; while a dozen or more debtors were removed from the ship by the bailiffs before she sailed.

The passage across the Atlantic was slow and tedious. The volunteers grumbled and quarrelled, and it was not surprising that by the time the *Two Friends* reached the Danish island of St Thomas, Charles had wearied of all but a few of them. Nor was the situation improved by the non-appearance of the accredited agent of the Independents, who was to further their passage from St Thomas to the mainland. Such revolutionary ardour as Charles may have had began to cool. Indeed for him, as for many others, service with the rebels was only a means to an end – a new life among the boundless opportunities offered by the developing American continent. Together with a score or so half-pay officers, he now accepted an offer from the American consul of a free passage in a schooner to Amelia Island, off the coast of Florida, where a Scottish adventurer named Sir Gregor McGregor was planning to invade the Spanish possessions of East and West Florida, with an army of American volunteers unofficially backed by the United States. But unknown to the consul at St Thomas, General McGregor had already left Amelia Island for a new command in Nassau; and at the end of October we find Charles not in Amelia Island, but in Charleston, South Carolina, having finally abandoned all ambitions to a revolutionary career, and set on finding agreeable civilian employment.

He did not have to seek long. On Christmas Day 1817 he wrote to his wife:

Know then that the state of South Carolina, intending to improve their internal Communications, have just selected a Gentleman here to the office of Civil Engineer, whose duties are to survey the Country, project and construct Canals, Improvements &c. &c . . . it was necessary that the officer thus appointed should have somebody to assist him, and just as this point was agitating I arrived from Amelia, and was recommended for that purpose. . . . This has been effected through my being the grandson of Dr Hutton, and an Irishman. . . .

The city of Charleston, where Charles Vignoles now found himself, was the capital of South Carolina, and owed its name and foundation to the time of Charles II. It was a place of some importance, able to boast a public library, bookshops, newspapers and a racecourse. It competed for the honour of having the first playhouse in the United States, and the beautifully proportioned houses, in which the great cotton planters lived, were inspired by the works of Wren, Gibbs and Inigo Jones. One can understand the immediate appeal of such a society to Vignoles's taste. Nor could he have arrived at a more fortunate moment. In the period of development following the War of Independence, the opening up of new cotton plantations, as well as the consolidation of the older ones, was giving rise to a demand for thorough surveys and accurate maps. In a colonial population military engineers were naturally qualified for such work. Major Wilson, Vignoles's new chief, was no doubt such a one, and a veteran of the English war. Vignoles's experience in this field, since the day when he mapped the Willemstadt fortifications with the aid of his feet and a pair of compasses, was scanty. His account to Mary of how he was appointed is tantalisingly incomplete. But we may surmise that his success was due in no small measure to his ebullient self-confidence, personal charm and some judicious name-dropping. The fact that Dr Hutton's name seems to have been a talisman in military and engineering circles even across the Atlantic underlines the cultural ties that still existed between the United States and the mother country. Moreover it is not impossible that Vignoles hinted at qualifications that he did not possess. At the same time, in an age of degrees and diplomas, it is hard for us to realise how many of the great engineering pioneers of the nineteenth century began their careers with no qualifications at all.

Vignoles's salary was to be 1000 dollars per annum, plus expenses, with permission to work in private practice as a surveyor, which might add a further 500 dollars. He calculated this at about £1 per week. While all Mary's talents as a manager would at first be necessary to make ends meet, he assured her that his income would undoubtedly increase, and that they would enjoy a situation in the best of Charleston society. He was determined she should join him as soon as possible.

Even before arriving in America he had hatched a plan with two of his fellow adventurers that Mary and their two wives should travel out together. A bargain was struck with the captain of a Liverpool-bound ship for a passage for the three ladies on his return to Charleston. Charles wrote long and detailed instructions to Mary, which he repeated in duplicate and mailed on different ships, in case of letters going astray. He listed a multitude of things she must bring with her, both for Charles and herself. His own needs included shirts, handkerchiefs, silk stockings, shoe buckles and a writing-desk; and she must go to Brookman & Langdon's in Great Russell Street and get

six Dozen of their very best HH *Sketching* Black Lead Pencils . . . such as they send to the R. M. College for the Fortifications Plans . . . moreover 20 sheets of that kind of *tracing paper* I got from Ackerman's just before I went to France . . . a plain box of Colors . . . a couple of the best Penknives London produces! a super excellent Razor Strop, and as much of the best Drawing Paper and Stationary for general use as you can afford to get.

Letters took anything up to ten weeks to arrive, which only added to the pains of separation. By the time she received Charles's first letters,

City of
Charleston,
South Carolina,
from the sea, *c.*
1820. Aquatint
by William
Keenan after a
drawing by
Charles Vignoles.
The inscription
states that the
drawing was 'in
the possession of
Henry Ogilby
Esq. His B.M.
Consul,
Charleston, So.
Ca.', but it has
not so far been
traced. I. N.
Phelps Stokes
Collection, New
York Public
Library.

Mary knew that she was pregnant (a possibility Charles had not over-looked, as a factor in deciding when she might be fit to travel), and while his letters are anxious but optimistic, full of plans for her journey and for their coming life together, hers grow more and more fearful at the prospect of a sea voyage with a young baby, to say nothing of the ordeal of childbirth itself, with no husband at hand to comfort her. Besides all this she was desperately short of money. She had had to give up her dressmaking, and the War Office had suspended Charles's half-pay. There were two reasons for this: a trifling error in the costing of the Royal Scots' meat ration at Fort William, and the assumption that he had taken a commission in Bolívar's army.

It was not until the beginning of May that Charles became fully aware of Mary's difficulties, and by this time, on 2 May 1818, a daughter had been born, after a long and difficult labour. She was christened Anna Hester, after two friends who stood as godparents, since Mary was too ill to give any instructions about her name, and her father's wishes were as yet unknown. Had Charles been there she would have been baptised Camilla, after his mother, and Camilla she was always called. Three months later mother and child and a nursemaid embarked at Gravesend for the voyage to Charleston. A friend who saw them off notes that there was a cow on board the ship, 'so very fortunate for the Babe'.

Charles's letters were full of plans for the future, but he said little to Mary about his own life at Charleston, beyond mentioning the names of his friends who would be pleased to welcome her. Among these were a number of young ladies who, he told Mary, could scarce believe he was

married. For a time he lived with an English couple named Dalton; the husband had lately kept a large chemist's shop in the Haymarket. But in the hot fever-ridden summer months he moved out to live on his own in a couple of rooms in the healthier atmosphere of Fort Moultrie, on an island commanding the harbour entrance. While watching for the arrival of the ship which was to bring his wife and child, he worked on the setting out and drawing of the maps of the ground surveyed earlier in the year. Partners in his work were two men of French extraction, probably members of the Charleston Huguenot community, with whom he would naturally have been in sympathy. One of these, Colonel Petitval, was to be his friend and colleague during all the time he spent in America; while the name of Henry Ravenal is linked with that of Vignoles on maps of the Beaufort and Charleston districts of South Carolina which still survive.[1] It is probable that it was during his stay at Fort Moultrie that Vignoles made the drawing of the Charleston water-front which is the basis of Wil-

Map of Charleston District, South Carolina, 1820. Surveyed by Charles Vignoles and Henry Ravenal. From *Atlas of South Carolina* by Robert Mills, engraved by H. S. Tanner, 1825. Copies are in the British Museum and in the Charleston County Library.

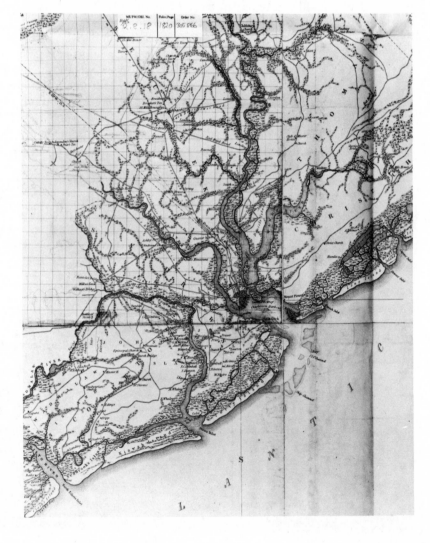

liam Keenan's attractive aquatint.

With Mary's arrival in Charleston, the correspondence is suspended for over a year, except for a few letters written early in 1820, when Charles was working up country on the Savannah River. These show that he had plenty of work, both public and private, but their tone suggests that their life together in America had not brought the couple the unalloyed happiness they had hoped for.

Despite Mary's prudence they were once again in difficulties over money, a subject on which she reproached him bitterly. A second child, Charles, had been born in August 1819, both children were suffering from the climate, and Mary was finding it hard to make ends meet. Charles, like many young men of his time, suffered from a congenital inability to live within his means; and the trouble was accentuated by the system of contracts under which he worked, whereby he was only paid for work when it was completed; and his clients kept him waiting months, even years, for payment. Even the state engineer's department was months behind with the payment of his salary.

In a mood of disillusion with the country and his prospects, he decided to return with his family to England for a few months, leaving the survey work in the hands of his partners. In July 1820 they were back in England, Charles hoping to clear up the matter of his half-pay and to reconcile himself with the Hutton family. He managed to persuade the War Office to restore his pay, together with the two years of arrears due; but though his aunt received him kindly, promising that she would always be a friend to his children, his grandfather still refused to have anything to do with him.

By the beginning of November Mary was once more pregnant, and whether by her choice or his, or because of their shortage of money, Charles decided to return to America alone, leaving Mary and the children established in simple country lodgings in Andover. Miss Hutton's offer of friendship does not seem to have extended to Charles's wife, but his solicitor agreed to keep an eye on her financial affairs. She would have his half-pay to live on, and he was confident that once back in Charleston he would be able to send her some of the money owing to him. There was no doubt that they both felt the separation deeply, but if Mary had misgivings about the future, Charles was as usual full of optimism.

Arriving back in Charleston, he found a warm welcome from Colonel Petitval and many old friends, and plenty of work on hand. Orders were coming in for plans of the city and the racecourse, and a number of important surveys for private individuals. By the end of March the last district map of South Carolina was finished. Since first arriving in America Vignoles had acquired considerable practical experience as a surveyor and draftsman. At the same time the hard field-work of the South Carolina surveys, involving long hours in the saddle or on foot in undeveloped country, had toughened him physically and developed the remarkable capacity for sustained and persevering effort which was to be a characteristic of his working life. Now, during the next three years, this experience was to be enriched and consolidated.

But although there was plenty of work, some clients were still in arrears with their accounts, and by the middle of the summer Vignoles was again short of money. With the coming of the July rains and the threat of fever, he once more left his partner, Petitval, to carry on their business in Charleston, while he went off to try his luck in East Florida, recently acquired by the United States by treaty from Spain. Here he secured the post of City Surveyor in St Augustine, with a view to making a map of the harbour and of new streets.[2] He was to receive no payment beyond expenses, but he foresaw many opportunities of private work offered in the new state. That he was not averse to trying any line appears from an entry in the *East Florida Gazette* for 22 December 1821, which records his appointment as Public Translator and Interpreter. (Since arriving in Charleston he had added Spanish to his linguistic accomplishments.)

Meanwhile he was distressed by the difficulties of providing for his wife and family, and the cost of maintaining two establishments. At times even his self-confidence began to waver and he contemplated returning home, but at the thought of Dr Hutton's 'flinty obstinacy' towards him he wished never to set foot in England again. Yet he was ambitious, and determined to be famous one day – but for what? Characteristically he now embarked on a new undertaking which, though it might one day bring him fame, had no promise of any immediate financial return. A member of the Land Claims Commission of Florida, realising the need for a new map of this latest addition to the Union, suggested that Vignoles should produce one, and that he should write a topographical memoir of the state to go with it. The U.S. military authorities may possibly have commissioned the survey, but nothing seems to have been said about the expenses of publication, which in the event devolved on Vignoles himself. However he had high hopes of the work bringing him in money as well as distinction, although he was mortgaging his time and skill for a payment in the distant future.

He set off in February 1822 in company with a U.S. general and his staff. The expedition was to cover the east and south-west coasts of the Florida peninsula, visit the islands of the Tortugas and the Florida Keys and return to St Augustine overland from Tampas Bay. At the end of March he was back in St Augustine, to be off again almost immediately for another two months in the interior of the peninsula. Shortly afterwards he had a narrow escape from drowning when thrown from his horse while fording a flooded river. He contracted a serious fever and was ill for a month, during which time he was distressed to hear from Mary that her baby, Thomas, the son he had never seen, had died of scarlet fever.

Knowing that the letter would take two or three months to reach her, he wrote hoping time had soothed her first anguish and that she would have found consolation for her loss in the presence of the other two children. He was more than ever resolved to stay in the United States, now that there was a prospect of success with the Florida map and pamphlet; though he told Mary that his resolve was tinged with regret at the absence of cultivated society, and the lack of polish of the Yankees.

By the beginning of September the map and pamphlet were completed and he was in New York looking for a publisher. To get there he had been obliged to borrow heavily, presumably on the security of orders for work in Florida not yet paid for. But the prospect of getting his work into print buoyed him up. Besides, in New York there was the possibility of further employment. There was talk of a canal between New York and Philadelphia, and of a survey of the Great Lakes. And before long he was boldly applying for the post of Principal Civil Engineer of the Board of Public Works of Virginia, supported by testimonials from John Geddes, the former Governor of South Carolina, and from the Principal Engineer of the City of Charleston – none other than Colonel Petitval himself! His long-suffering partner, who seems to have combined a genuine affection for Vignoles with a shrewd understanding of his character, doubted whether anything would come of his schemes; and urged him to return to work on a revision of their city map which had just been accepted by the authorities.

Observations upon the Floridas, by Charles B. Vignoles, was published by Bliss & White of New York at the beginning of March 1823.[3] It was an octavo volume of 153 pages with 40 pages of appendices. At the same time appeared the new map of Florida, engraved by H. S. Tanner of Philadelphia. This was either 'done up with the Observations' or separately as required: 'Price 2 Dollars plain; 2.50 Dollars colored: 4 Dollars, mounted on rollers and varnished or done up in portable form.'

The book provides a brief survey of such topics as the history, topography, soil, 'appropriate articles of culture' and climate of the newly acquired territory, with separate sections on the Florida Keys, the Indians and land titles. The style is factual, and by modern standards somewhat prosy. The author himself seems to be aware of this, when he writes in the introduction: 'Those who may peruse these pages must not expect the glowing narrative of an agreeable excursion, through regions comparable to a paradise . . . [it may seem] dry and tedious to all not immediately interested in the resources of the territory.' Vignoles does not attempt to disguise the fact that much of his material is second-hand and honestly acknowledges his sources. He admits that he gives most attention to the Atlantic coast, which he knows best, and that although the facts about the coastal belt are based on his own actual surveys, his acquaintance with the interior is incomplete. Moreover West Florida is hardly touched on, for the reason given that it was impracticable to get there. The author is content to say that former accounts show the soil and climate to be not different from those of Mississippi and Alabama.

The book is thus by no means perfect, but may well have fulfilled the purpose of a guide to intending settlers, which Vignoles hoped it would become. 'Sensible of all possible respect for the opinion of an enlightened public, the work is offered to them, with all its imperfections on its head.' It has not been possible to trace a copy of Vignoles's map, but there exists a map of Florida, engraved by H.S. Tanner and dated 1827, which may well be based on it, though no surveyor's name is recorded.[4]

Vignoles was now thinking seriously about settling in the U.S.A. as an American citizen. He was aware of a family tradition, to which there is a reference in Dr Hutton's papers, that a grant of land in Florida had been made to his grandfather Francis Vignoles, whose regiment had been stationed there, at the time of the British rule in that state. If he could establish his title to an estate he would be well on the way to enjoying the rights of citizenship. But a new factor now entered the situation. An obituary notice appeared in the New York press from which Vignoles learned that Dr Hutton had died on 27 January 1823. His grandson had for so long given up any hope of a reconciliation that he was comparatively unmoved at the news. He did not suppose that he had been left a shilling. He was the more determined to abandon England for ever.

Yet within two months he had once more changed his mind. On 13 April he received a letter from Mary in which she confirmed that the Doctor had remained inexorable to the last, but added that his and Miss Hutton's solicitors were much concerned about the injustice done to Charles by the terms of his grandfather's will. Mary advised that he should lose no time in writing a letter of condolence to his aunt. If he had but a year's income in advance it might be politic to return to England. Otherwise she did not think it prudent. But counsels of prudence, especially those coming from his wife, were calculated to have opposite effects from those intended. On 21 April Charles booked a passage to Liverpool.

The reasons for this sudden decision are not far to seek. He was completely devoid of money. There was little hope of early profit from his book. He had been unsuccessful in his ambitions for an appointment in the North, and the hope of collecting the money still owed to him in the South seemed as remote as ever, for business in South Carolina was badly affected by a world slump in cotton. And although the Congress delegate for Florida had nominated him for the post of Surveyor General in East Florida, he was discouraging about Vignoles's claim to an estate.

But a letter from Colonel Petitval indicates what was perhaps the most telling reason of all. The unfortunate Vignoles still owed so much money in Charleston that Petitval feared he might be arrested if he returned there. 'Les vampires redoublent de voracité', he wrote, 'depuis qu'ils savent que vous êtes nommé arpenteur général de la Floride de l'est.' He bore him no grudge for his decision to return to England. Indeed he had expected it as soon as he heard of Dr Hutton's death. He knew Vignoles well enough to realise that once the 'flinty obstinacy' of his grandfather was removed, he could hope for a reconciliation with his aunt, and all that that would entail in the way of financial support. Petitval did not rule out the possibility of his partner returning to America. But he advised him not to attempt it without having raised sufficient money to pay his debts in advance. Only then would he be able to set foot in Charleston again, 'tête levée'.

In fact he never did. It would be interesting to know what was the final balance between his debts and the money he was owed.

3 The Liverpool & Manchester Railway, 1823–1827

The England to which Charles Vignoles returned in 1823 was a country of striking contrasts. Since the end of the Napoleonic War, the masters of the mills and factories of the north, enriched by the unique trading position enjoyed by Britain during its closing years, had been able to ride out a succession of slumps. Not so the mass of cheap unorganised labour by whose efforts they had made their wealth. In the country, the enclosure acts had enriched the landed gentry at the expense of the country people. Whether in factory or on farm, the rates of pay were dictated from above and fluctuated violently with the ups and downs of trade. Thousands of discharged soldiers and sailors had flooded the labour market, thereby forcing up prices and creating unemployment. Hunger, resentment and despair spread unrest among the workers, resulting in clashes with the authorities of which the so-called 'massacre of Peterloo', in August 1819, had been the most serious example.

Yet England was still one of the richest countries in the world. The aristocracy lived in conditions of unbelievable luxury, their manners and way of life being aped by a rising middle class. Much of their wealth was pouring into the building and engineering schemes arising from the rapid expansion of industry and the consequent swift growth in population. In London whole new quarters were being built: Nash's Regent Street and the terraces of Regent's Park, as well as dwellings for humbler folk. John Rennie's Waterloo Bridge had been followed by the iron bridge at Southwark, which boasted the largest cast-iron arch ever constructed; and the younger John Rennie was about to begin the building of the new London Bridge, designed by his father before his death in 1822.

Further downstream, at Rotherhithe, Marc Isambard Brunel was canvassing support for his Thames Tunnel. Meanwhile, away in the north, Thomas Telford had just finished the Caledonian Canal, and was forming the last link in his London to Holyhead coach road with the Menai Straits suspension bridge. And just as transport by road and canal was reaching a peak of development, a new engineering phenomenon had made its appearance.

It was in 1823 that George Stephenson and the Quaker Edward Pease established the first locomotive engine works, in Newcastle. In the same year the construction of the Stockton & Darlington Railway was begun, with Stephenson as its engineer. On this railway, passenger trains were to be hauled for the first time. On the other side of the country, the railway promoter William James, whose schemes extended to places as far apart as Moreton-in-the-Marsh and Whitstable, had organised the first surveys for a railway from Liverpool to Manchester, one of his assistants being Stephenson's son Robert. The railway age had begun, an age which was to offer boundless opportunities to men endowed with brains, enthusiasm, energy and a true pioneering spirit. Charles Vignoles had all of these; and soon time was to show how wise he had been to return to England.

First he had to regain the favour (and possible financial support) of his aunt, Isabella Hutton. With this in mind, after spending a few days with his wife and children, who were in lodgings at Littlehampton, he presented himself at Bedford Row. Miss Hutton was dining with her brother, now a general on half-pay. The latter refused to see his nephew, recommending that he should return to America at once. At first Miss Hutton was of the same opinion, but before long she and Charles were reconciled, though her favour did not at first extend to receiving Mary in her house. It was some weeks before she relented sufficiently to allow Charles's family to occupy three rooms at 34 Bedford Row while he sought suitable lodgings for them elsewhere. It was understood that Miss Hutton would only see the children when it pleased her to send for them. Relations with Mrs Vignoles remained cool.

Once established with his aunt, Vignoles set about looking for employment. He was confident that Dr Hutton's friends would help him into the world of engineering. Meanwhile he embarked on the survey of a gentleman's estate at Brentwood Hall, Essex, for which he was paid 60 guineas. He also found means of earning money with his pen. This was a contract to contribute some geographical articles to Smedley's *Encyclopaedia Metropolitana*, a periodical publication which had recently reached the letter C, enabling Vignoles to offer information on such places as Charleston, Chesapeake Bay and the Chattahootchie River. Finally, towards the end of the year, backed by testimonials from Dr Olinthus Gregory, who had succeeded Dr Hutton at the Royal Military Academy, and from his old tutor Thomas Leybourn, Vignoles obtained his first engineering post, as assistant to James Walker, at that time chief engineer to the London Commercial Docks.

Walker was planning the erection of a suspension bridge, designed by Captain Samuel Brown, R.N., which was intended to link Telford's St Katherine's Dock (then under construction) with a proposed South London Docks in Bermondsey. Evidently Mr Walker was impressed by his new assistant's skill as a draftsman, for he entrusted him with the final drawings of the bridge, and Vignoles records in his diary that on 23 February 1824 he waited with Captain Brown on the Duke of Wellington, to present the plans for his approval.

Other work soon followed. Possibly on Walker's recommendation, Vignoles was engaged by Tierney Clarke to work on the drawings of the first suspension bridge at Hammersmith, and by Joseph Yallowlee to assist with those of Meux's new brewery at Barnes. In such spare time as he had left he continued his contributions to Smedley's encyclopaedia.

His diary records that he worked long hours, frequently seven days a week, and was paid in arrears for the number of days' work done.

By the middle of 1824, he was fairly established in London, with an office in Hatton Garden, three pupil-assistants of his own, and a home on the first floor of a house in Kentish Town. A week before his thirty-first birthday he dined with the newly-formed Institution of Civil Engineers at the London Coffee House. Mr Telford, to whom he was introduced by Dr Gregory, was in the chair. The doyen of the engineering world must have appealed to Vignoles, for he shared the younger man's enthusiasm for literature. Telford had shown some interest in the map of Florida, which Vignoles was hoping to sell in London. For he had not entirely turned his back on America. Colonel Petitval wrote hoping to enlist his help in producing a guide to the principal cities and roads of the United States. In August Vignoles was invited by his solicitor, Charles Hanson,[1] to go over to New York on business connected with the estate of one of his clients. Considering the terms on which he declared himself prepared to go (a fee of a thousand guineas and expenses) it is not surprising that the invitation was withdrawn. Perhaps he recalled Petitval's advice to him on his leaving America. But the figure suggests that he was well satisfied with his prospects in England.

The following year he was to have one final look across the Atlantic, when for a short time he was negotiating with a company proposing to develop road and canal communications in Mexico. This proposal was to include a canal through the Isthmus of Tehuantepec, which if built would have anticipated the Panama Canal by a quarter of a century.

A temporary partnership with Joseph Yallowlee as engineer to a company proposing to build a coaling dock in the Isle of Dogs proved to be abortive, but early in 1825 Vignoles was again employed by Walker, first on a survey for the Cheltenham Water Works, and then in the fen country. (An interesting expense account for the Cheltenham trip survives, amounting to £20 9s 2d for the week, including coach fares, £3 10s 0d return each for himself and his assistant Charles Forth, board and lodging £6 10s 5d, servants £1 1s 0d, tips to guards and coachmen 12s 0d each journey, and wages for the chain carriers for the week 10s 0d. Additional to these were the cost of materials for the drawing, tracing, and mounting the map, and the coach hire to and from Vignoles's lodgings and Mr Walker's office.)

The week in Cheltenham provided a good example of the intensity with which he was prepared to work. He travelled down by the night mail one Sunday, and returned the following Sunday, after completing a week surveying by day and drawing by night. A diary entry reads:

Thursday 3rd. March. Rose this morning at ½ past 5 a.m. and began to fill in the New Buildings by actual Measurement: but finding the map so extremely erroneous considered it best to sketch in the additions. Employed sketching till 6 p.m. when we returned and continued till midnight in fixing same on Map.

After a week's respite in London, during which the clerks completed copies of the Cheltenham map, and Vignoles compiled and sent off another batch of articles to Mr Smedley, ranging from Crane and Creole to Cuba, Cumberland and Curaçao, he was off to Norfolk. In 1822 John Rennie and Telford, engineers to the Ouse and Bedford River, had recommended enlargement of the Eau Brink Cut, an artificial channel completed by them two years before to improve the river mouth between King's Lynn and the Wash. James Walker had undertaken the contract, and instructed Vignoles to carry out a complete survey of the river and its tributary dykes. The survey took five weeks, much of it working from boats, and involved the compilation and verification of much tidal data. To one who had learned in the hard way on the Florida and South Carolina coasts it may not have presented many difficulties, but the March winds over the Fens must have been as trying as the heat of the tropics.

Hitherto all Vignoles's work had been concerned with canals, docks or bridges; but in the last week of May, having completed his work for James Walker on the Fens survey, he was sent for by John Rennie, who with his brother George had been commissioned to survey possible routes for a railway from London to Brighton. The Rennies were busy with other projects, notably London Bridge, and they seem to have had no hesitation in entrusting one of these surveys to Vignoles, now well known to them through his work for Walker.

This was altogether a new type of work for Vignoles, but no doubt he embarked on it with his usual confidence. Travelling by horse-chaise, and sleeping at hostelries in towns and villages on the way, he explored and levelled the whole of the proposed line in just under four weeks. His task was to choose the most favourable line from the point of view of gradients, and calculate and plot a vertical profile of the whole line. This time three assistants went with him and he was driven by his own coachman. Chain carriers were engaged casually from day to day. Between Leatherhead and Mickleham he records that he was 'much annoyed by landed proprietors', a well-known hazard for railway surveyors. Otherwise the survey, which followed the course of lines to be built many years later, seems to have gone smoothly.

The association with the Rennie brothers marked a turning point in Vignoles's career. Not only did it introduce him to the field of engineering in which he was to be engaged for most of his life, but also to the area where most of his work was to be centred. It happened that the directors of the Liverpool & Manchester Railway, having failed to get their bill through Parliament in the spring of 1825, had dismissed their engineer George Stephenson and appointed John and George Rennie in his place. While Vignoles was still working on the plans and estimates of the Surrey and Sussex line, John Rennie asked him whether he would be prepared to leave in a few days for the north, to carry out a new survey of the Liverpool & Manchester line. Vignoles agreed. And on the evening of Wednesday 13 July he left London for Liverpool, the base from which he was to operate for the next 15 years or more of his life.

The rejection of the first Parliamentary bill to authorise the construction of the Liverpool & Manchester Railway had been brought about by the violent opposition of the landowners and the canal proprietors, whose monopoly of transport between the two cities the railway scheme was designed to break. George Stephenson's survey of the line had been made with some difficulty, as landowners turned out their farmers and gamekeepers in force to bar his entry onto their land. When he resorted to surveying by moonlight, the Duke of Bridgewater's manager, Captain Bradshaw, had guns fired over his grounds to discourage the surveyors.

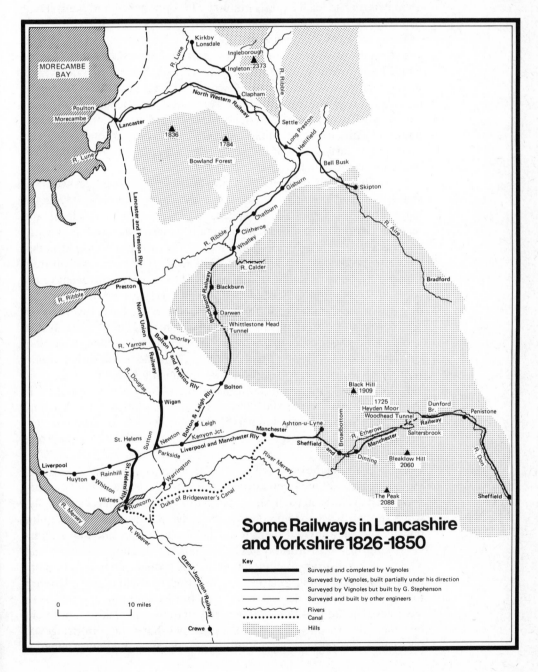

Some Railways in Lancashire and Yorkshire 1826-1850

Key

───────	Surveyed and completed by Vignoles
─────	Surveyed by Vignoles, built partially under his direction
───	Surveyed by Vignoles but built by G. Stephenson
─ ─ ─	Surveyed and built by other engineers
∿∿∿	Rivers
••••••	Canal
░░░	Hills

0 10 miles

At the same time, owing to his own preoccupation with the Stockton & Darlington line, Stephenson had relied heavily on his assistants and had not himself checked all their figures. It was hardly surprising that the opposition, who had briefed the most skilful of counsel, and enlisted a number of competent engineers as witnesses for the Parliamentary enquiry, were able to show that Stephenson's plans had serious defects. In cross-examination the self-educated Northumbrian engineer had proved no match for the legal experts put up by his opponents, and the Bill had been rejected, at the end of May 1825.

These were the circumstances in which the Rennies had been charged with the task of making a new survey of the line, and preparing plans for a new Parliamentary Bill. John Rennie writes in his autobiography: 'As we were left entirely to our own discretion to adopt the old or to choose an entirely new line we selected the present energetic and talented engineer, Mr Charles Vignolles [sic] to make the necessary surveys for Parliament.' This suggests that although George Rennie accompanied Vignoles in the early stages of the survey, the actual choice of the line may well have rested with the latter.

According to his diary for July and August 1825, Vignoles was constantly engaged in going over the line with George Rennie, studying the plans of William James's and Stephenson's surveys, and taking levels and sections. By the beginning of September he had seven assistants at work, and his establishment was completed by his own phaeton and horse, which, with his coachman and heavy baggage, had been sent up from London by canal. The line he ultimately chose passed somewhat to the south of Stephenson's, avoided the town of St Helens and the properties of some of the more vigorous opponents, and was to enter Liverpool directly from the east by means of a deep cutting at Olive Mount. It thus avoided the long detour made by Stephenson in approaching Liverpool from the north via Bootle. In order to reach the dockside without interfering with the city streets, a tunnel was planned on a steep incline from Edgehill to Wapping. At its eastern end the railway was to stop short in Salford without entering Manchester. The line was thus shorter and more direct than Stephenson's, but was still to pass over the peat bog of Chat Moss, an undertaking on which scorn had been poured by Stephenson's opponents.

Vignoles found the opposition of the landowners in no way diminished, though he and Rennie fared rather better with Bradshaw, whom they encountered one day on Chat Moss. The latter greeted them very cordially, while stating that he would oppose the railway with every means in his power, and then invited them to his house. In his presidential address years later to the Institution of Civil Engineers, Vignoles declared that while making a survey by moonlight he was brought up before Captain Bradshaw for poaching and trespass. 'Mr Bradshaw had contrived to earn for himself a terrible name for severity, but I found him a gentleman.' Vignoles claimed that this interview sowed the seeds of an arrangement by which the Marquis of Stafford, a principal shareholder in Bridgewater's canal, gave up his opposition to the railway and demonstrated his support by the purchase of a thousand shares in the company.

On 26 December the new prospectus for the railway was issued. It aimed to forestall opposition by deliberately soft-pedalling the possibility of using locomotives, rather than horses or stationary engines and ropes, and emphasised the care taken to avoid game preserves, and the avoidance of interference with the Liverpool streets made possible by the Edgehill Tunnel.

The new Parliamentary Bill had its first reading on 7 February 1826. No official record of the proceedings of the Commons committee exists, but from other accounts (notably those of R. E. Carlson and L. T. C. Rolt) the Bill seems to have had a relatively easy passage. In the Lords committee, which opened on 13 April, it became clear that the changes in the plan had taken a good deal of the heat out of the attack. Moreover George Rennie and Vignoles, on whom the brunt of the cross-examination fell, were fully armed and ready to defend their plans. Vignoles showed up particularly well, even when pressed hard by the formidable counsel Alderson, on the dangers which the boring of the tunnel might afford to Liverpool's water supplies. There can be no doubt that Vignoles owed much of his success to his early legal training, which was to help him through many subsequent Parliamentary committee proceedings, and to cause him to be much in demand in later years as a Parliamentary witness. He was also brimming over with enthusiasm and self-confidence. In the words of Dendy Marshall: 'The plans were now unassailable. Vignoles had not surveyed large tracts of land in America for nothing.' And O. S. Nock emphasises that 'there was all the difference in the world between the bearing and speech of this polished young soldier, and "Geordie the engine-wright" '. In the end only Lords Derby and Wilton voted against the Bill in the Lords committee, and on 5 May, after its third reading in the Lords, it received the Royal Assent.

Having obtained their bill, the railway Board had now to decide who was to build the line. John Rennie writes: 'we naturally expected to be appointed as executive engineers, after having with so much labour and anxiety carried the Bill through Parliament'. But the Board had other ideas. Whatever Stephenson's shortcomings might be as a Parliamentary witness, they still had faith in him as a practical engineer; and they had some doubts as to whether John and George Rennie, with their many other commitments, would be able to give the railway the attention it required. At a meeting held on 5 June they resolved to offer George Rennie the post of consulting engineer, and to appoint either Stephenson or J. U. Rastrick as resident under his general direction. At the same time they took the unusual step of inviting Vignoles to come immediately to Liverpool to begin staking out the line. For this he was to be paid four guineas a day without expenses. Having no further employment with the Rennies, he accepted. But it was made clear to him that though he was now engaged directly by the Company on what was virtually the first step in building the line, they had not yet decided on either a principal or resident engineer.

The Rennies' reply, delivered in person by George Rennie to a meet-

ing of the Board, was that they would agree to undertake the superinten-
dance of the works, but that they insisted on having the right to appoint
their own executive engineer. Under no circumstances would they agree
to be associated with Rastrick or Stephenson since they were not members of the 'Society of Engineers' (later the Institution of Civil Engineers), though they would have no objection to Stephenson being entrusted with the locomotive department, which they considered should
be distinct from the other works.

On 21 June the Board decided that the Rennies' terms were unacceptable. The brothers withdrew, feeling that the Directors had 'behaved extremely ill' towards them, and early in July, Rastrick's nomination having
been withdrawn, George Stephenson was appointed chief engineer, with
Josias Jessop as consultant. But what of Vignoles, who had now been laying out the line for over three weeks? Chosen by the Rennies to survey
the line and to defend it in Parliament, and directly engaged by the Company to start laying it out, he was in a strong position, but one not without
its delicacy. It was almost certain that if the Rennies had had their way
he would have been their choice for the post of resident engineer. For
nearly twelve months he had been the principal engineer on the line, and
the man who knew it best, having selected it and surveyed it himself. He
had every right to be retained as assistant to Stephenson – which was what
the Board proposed he should be. But the fact that he had succeeded
where Stephenson had failed, and that Stephenson was now appointed
to build the line as modified by Vignoles and George Rennie, was hardly
likely to endear him to his chief. The situation was not improved by the
extreme contrast between their characters. Stephenson was a rough dia-
mond, self-educated, slow in speech, but intensely practical and something of a mechanical genius; while Vignoles was volatile, quick-
tempered, cultivated, imaginative, a highly-experienced surveyor, but
lacking in mechanical knowledge.

It says a great deal for both men that in spite of all this they were prepared to give the arrangement a trial. Stephenson accepted Vignoles, but
subsequently engaged one of his pupils, Joseph Locke, as a further assistant. Vignoles, on his side, found encouragement from Miss Hutton. On
hearing of his plight she enlisted the help of Mr Edward Riddle, professor
of mathematics at the Greenwich Hospital Naval School. On 26 July Riddle wrote to Vignoles, enclosing an open letter of recommendation to
Stephenson. To Vignoles he wrote: 'I know Mr Stephenson very well. His
talents . . . are altogether of a practical nature; his theoretical knowledge is slight, but he seems to have a sort of intuitive faculty for unravel-
ling the complications of machinery, and overcoming difficulties of simple contrivances.'

But such an uneasy relationship was doomed to failure. It was not long
before Stephenson was at odds with Jessop, whose appointment as consultant was a source of further resentment. Jessop disagreed with
Stephenson on many points, and Vignoles was inclined to agree with Jessop. This can only have further identified Vignoles in Stephenson's mind
with the clever men who had discomfited him in Parliament, and with the
engineers who had refused to work with him because he was not one of

their Society. Even Riddle's letter of recommendation was a source of offence, since Stephenson interpreted it as describing Vignoles as his partner rather than his assistant.[2]

Matters came to a head in November when Stephenson reported to the Directors that there was an error of 13 ft in the position of the workings of the Edgehill Tunnel. Vignoles had completed the survey for this in September, and since then four shafts had been sunk, and pilot boring begun. Stephenson put the blame for this on his assistant, and disclaimed any further responsibility for his actions. The accusation was repeated in a letter to Riddle, to whose recommendation of Vignoles he had hitherto not condescended to reply:

I am much obliged to you for complimenting me on my appointment as engineer to this concern. You also congratulate me on the able assistance I should meet with from Vignoles but I am sorry to say I have been strongly censured by the company for the blunders he has made in the works I put under his hands. He appeared to me to be a clever active man, but on trial I find his surveying is by no means to be depended upon . . . I sincerely feel for him, but I must feel for myself first. He was only put under me as a second assistant, but this work I put under his care, as I understood it was quite in his way.[3]

Vignoles, while admitting to a mistake in his initial survey, denied that he alone was responsible for the errors made in fixing the line of the tunnel. He had made the survey with the utmost care, but had left it to one of Stephenson's assistants to do the 'poling out' while he was absent working at the other end of the line. Had the 'poling out' been properly done the error would have been discovered at once. Perhaps Stephenson remembered that he too had relied too much upon his assistants in his first survey; but whatever the rights and wrongs of the case, he was evidently determined to make Vignoles the scapegoat, and the Directors could hardly fail to support their Chief Engineer, nor overlook the fact that Vignoles no longer enjoyed his confidence. In private they expressed their regret that 'the queer temper of Mr Stephenson' would not allow Vignoles to remain in Liverpool, and left it to him to send in his resignation when it was convenient to him to do so. This he finally did on 2 February 1827. As Josias Jessop had died suddenly the previous October, and had not been replaced as consultant, Stephenson was now free to appoint his own assistants and manage affairs in his own way.

A draft copy of a letter from Vignoles to Riddle, dated 14 January 1827, gives his view of the affair. He writes:

I do also acknowledge having on many occasions differed with him, because I . . . consider (and that too in common with almost all other engineers) that the mode Mr Stephenson proposes to put the works into the course of operation is not the most eligible; and because it appears to me he does not look on the concern with a liberal and expanded view; but considers it with a microscopic eye, magnifying the importance of details . . . pursuing a petty system of parsimony, very proper in private Collieries or small undertakings but wholly inapplicable to this National work . . . I also plead guilty to having neglected to court Mr Stephenson's favor by sycophantic expressions of praise . . . or by crying down all other engineers, particularly those in London. . . . All these circumstances gave rise to a feeling of Ill-will towards me in Mr Stephenson's mind which he displayed on every occasion, particularly when I showed a want of practical knowledge of unimportant minutiae, rendered familiar to him by experience.

After giving his explanations of how the error in the line of the tunnel arose, Vignoles goes on to point out that Stephenson had never been satisfied with the course of the tunnel, and that he had seized the opportunity to persuade the Directors to shift the whole line of the tunnel by 40 feet to the northward, thereby making Vignoles's error appear greater than it was. The latter was justly annoyed that Stephenson, although already aware of the errors in the tunnel, had encouraged him to go ahead with plans to move his family up north from London. As a result he had ordered his furniture to be shipped round by sea, and incurred an unnecessary expense of £100. However, as he wrote to Riddle: 'The liberality of the Directors will indemnify me for this loss, and I trust I may quote this as a Set Off against Stephenson's libels, for had I given any solid or serious Cause of Complaint or reasons for Dismissal, I should not be repaid an Expense like this.' He also contested Stephenson's statement that he was his 'second assistant': 'Two assistant engineers were nominated by the Directors. Mr Locke (Stephenson's articled clerk) and myself. Our salaries were equal – but my name stands first in the Record of Appointment.'

Relations between Vignoles and the Board had always been cordial. With his formal letter of resignation he enclosed a private letter to Charles Lawrence, the Chairman, which concludes:

I must trespass for one moment more to say that from his first coming here I have ever been treated by Mr Stephenson with a degree of distrust and supercilious bearing in professional matters extremely painful to my feelings, and extending in some points to a degree almost ridiculous; the expectation of being relievĕd from this mortification constitutes the only consolation I can find against the many others I feel in vacating my situation.

Resignation from the post of assistant engineer did not end Vignoles's interest in the Liverpool & Manchester Railway, as we shall see later. It would be an exaggeration to say that if Vignoles had never existed the way would not ultimately have been built. But the part he played in its launching was crucial at the time. A Liverpool friend wrote: 'It would never have been a Rail Way but for you.' Before he left Liverpool the Directors passed a generous vote of confidence in his ability, which was some indication of his reputation in the city. He had no intention of departing under a cloud, and at this very moment was discussing with some of his business friends the idea of a tunnel under the Mersey. In a long letter to the *Liverpool Albion* dated 6 March 1827, he claimed that a tunnel would be of immense benefit not only in opening up road and rail communications with the south, but also to the agricultural trade between the Wirral and the city. He wrote 'as one who though a stranger in the Town is, from feelings of gratitude for favors conferred by many of its citizens, deeply interested in its welfare'. At the same time he refrained from suggesting who might be its engineer.

Having cast this pebble into the waters of the Mersey, where it lay undisturbed on the bottom for 60 years, Vignoles left shortly afterwards for London, where on 10 April he was elected a member of the Institution of Civil Engineers. His gratification in receiving this distinction would perhaps have been increased had he been able to foresee that it was one to which George Stephenson was never to attain.

4 The struggle for recognition, 1827–1832

It is not altogether surprising that Vignoles should have taken such pains to justify himself in his letters to Riddle and Charles Lawrence. The dispute with Stephenson had been a humiliating and disappointing setback to his hopes. The fact that the accusation of carelessness was not wholly without foundation only made matters worse. As it was through men like Riddle, Dr Gregory and other friends of his grandfather that he owed his entry into the engineering profession, it was important to him not to lose their confidence so early in his career.

He also had to justify himself to his wife. Mary's disapproval of Charles's lack of prudence had not diminished since their marriage. Ever since the Rennies had left Liverpool she had had grave misgivings about her husband's prospects with the L.&M.R., and in her role of critic and mentor she had not failed to tell him so. Now she hoped that the breach with Stephenson would be a lesson to him. She was right, in so far as the affair of the tunnel had at least taught Vignoles the importance of checking his own calculations, a point he was to be most meticulous about in years to come. But Mary's criticism only turned the knife in the wound, and made relations between them more difficult. The result was to confirm Charles in his determination to be master in his own house, while giving all his attention to a profession in which success was only to be won by unrelenting individual effort against relentless competition.

But before he left Liverpool an event occurred which must have done something to restore his self-esteem. Earlier in the year Marc Brunel had found time, amidst his preoccupations with the Thames Tunnel, to design a floating low-water landing-stage for foot passengers on the Liverpool dockside. In November his son Isambard, then 20 years old, visited Liverpool in connection with this project. He was shown over the railway works by Stephenson and Vignoles, and no doubt listened with the sympathetic ear of youth to the younger engineer's story. It happened that the elder Brunel was at this time looking for a new resident engineer for the tunnel works. He wished to appoint Isambard, but feared there would be objections from the Directors to the appointment of his own son, and at so early an age. Hearing from Isambard that Vignoles was leaving Liverpool, Brunel wrote to offer him the post. Vignoles accepted, saying that he would be free to start work on 1 February. In the meantime, however, Brunel found that he had been mistaken about the Directors' possible objections, and on 29 December he wrote again to Vignoles to cancel the offer, though adding: 'You can be assured that I could not have applied to anyone else, from the reliance I had on your qualifications for so arduous an undertaking.' He also invited Vignoles to visit the tunnel when he came to London.

Disappointed as he must have been, Vignoles replied: 'I beg you will believe me sincere when I assure you that I am truly rejoiced to find the Directors of the Thames Tunnel have had the prudence to appoint . . . an engineer as competent as your Son, for whom I fear I should have been an inadequate substitute.' This contact with the Brunels was to have important consequences.[1]

An engagement to survey some newly-acquired government land now took Vignoles to the Isle of Man. This may have been the result of an application he had made the previous year, backed by his uncle, General Hutton, for work with the Ordnance Survey. The assignment was substantial enough for him to settle Mary and his two younger children, Hutton and Isabella (the elder two, Camilla and Charles, had been sent to a boarding school in London), in a house in Douglas. Here Mary was shortly to give birth to another son, Henry.

The survey kept him busy for most of the year, but in November he heard again from Marc Brunel, this time with an invitation to relieve him of some of the detailed work on a project to straighten a stretch of the Oxford Canal near Coventry. Vignoles, always ready for new experience, left his assistants at work on the island, and returned to London to go over the plans with Brunel. While he was there Isambard conducted him on the promised visit to the Thames Tunnel.

Excellent relations were established between the elder Brunel and his new assistant, so good in fact that when in May 1828 the Canal Committee expressed a wish to employ Vignoles as chief engineer on his own, owing to Brunel's heavy commitment with his other work, the latter immediately agreed. In fact, he must have been only to glad to be relieved of the responsibility. In January the waters of the Thames had for the second time broken forcibly into the tunnel face, bringing death to six miners, and seriously injuring Brunel's son. The work of repair was arduous, the company's money was running short, and public confidence in the undertaking was shaken. In August all further work was suspended and the end of the tunnel had to be bricked up.

A number of the Tunnel Company directors, led by their chairman, William Smith, openly laid the blame for their misfortunes on Brunel. They encouraged suggestions for completing the tunnel from other engineers, and the project became the subject of public argument and discussion.

By the end of the year Vignoles was again in London, working with his assistants in his new office at 14 Furnivals Inn on the Parliamentary plans of the canal improvements. In London the talk was all of the future of the Thames Tunnel. It was inevitable that Vignoles should have seen here an opportunity for a major engineering work in his own right.

His intervention has been seriously criticised by Richard Beamish, in his *Memoir of the Life of Sir Marc Isambard Brunel*; and indeed, considering the friendly treatment he had received from the older man, his attempt to supersede him appears the height of ingratitude. On the other hand, it is only fair to point out that the Chairman of the Tunnel Company was set on replacing Brunel by another engineer. It is possible that Vignoles may even have persuaded himself that Brunel's offer of the post of engineer was sufficient justification for him to apply to complete the work. Yet he had only brief experience of tunnelling, and none of tunnelling under a river. Lack of experience however was the last thing to deter him, though he was never afraid of picking other men's brains. In this instance he made sure that the contractors he was to recommend for the work, Messrs Pritchard & Hoof, had been involved in the construction of several canal tunnels.

A note in Brunel's diary for 12 February 1829: 'Wrote to Mr Vignoles that nothing on his part, in his relation to having superseded me in the Oxford Canal business, has diminished the favourable opinion I entertained of him before', suggests that Vignoles was sensitive about his relationship with Brunel, even though he was about to attempt to supersede him in a much bigger undertaking. A key to his state of mind at this time may be found in his correspondence with Mary. She was once again pregnant, and worn out nursing a houseful of children laid low by an epidemic of measles. The tone of her letters veered from affection to bitter complaints. Vignoles himself was racked by a feverish sore throat and scarcely able to drag himself to the House of Commons proceedings on the Oxford Canal Bill. To add to his anxieties, young Charles had been suffering for some time from strange attacks resembling epileptic fits, and to crown all, at the beginning of March, the four-year-old Isabella died of the after-effects of measles. Vignoles wrote entreating Mary to be calm for the sake of their family and the unborn child: 'If it were possible I should leave London to-night, but this I cannot do! We are in the midst of all our Parliamentary Bustle, and I am in a most solemn treaty with the Thames Tunnel Company.' On 14 March he wrote again saying it was impossible for him to leave: 'I have a solemn appointment with the Directors . . . the result of which will be my appointment to conduct that great work instead of Mr Brunel.' Even when Mary fell seriously ill after the birth on 24 March of their son Olinthus, he still could not bring himself to leave London. To him the possibility of his appointment in place of Brunel had become a matter of greater importance than the health and happiness of his wife and children.

Early in April his first proposals for completing the tunnel were passed to Marc Brunel for his comments. Vignoles had insisted that this should be done before the proposals were put to the proprietors, 'inasmuch as the proposer [Vignoles was at this stage treating with the Board in secret, and his name did not appear on the paper] has a high regard and esteem for Mr Brunel both as a man of Science and as a Gentleman . . . and because he wishes to avoid the imputation of having acted either unhandsomely or disingenuously with respect to Mr Brunel.' *Qui s'excuse, s'accuse*, some might say.

Brunel's restrained comment on 14 April: 'We suspected Vignoles as the proposer of new terms', seems to indicate a good understanding of his former assistant's character, while saying much for his own tolerance. But on 20 April he criticised the proposals severely, particularly those which would allow the proposer to make certain unspecified experiments, before agreeing with the Company to have the works completed by contractors chosen by him, according to his own plans, so far unspecified. Brunel maintained that the Directors were buying a pig in a poke, and protested against 'principles pregnant with mischief to the work and ruin to the Company'.

There followed a period of bitter argument and discussion. William Smith, the company chairman, was not disposed to pay attention to Brunel's protests. He tried to persuade the shareholders, and the public, that the Company's only chance of success – which would depend on a government loan – was to abandon Brunel and seek alternative plans. The fact that Brunel was known to be a friend of the Duke of Wellington, and that he was quietly pursuing an application for government aid which he was confident would be eventually forthcoming, meant nothing to Smith. Indeed, so intense was his dislike of Brunel that he would have preferred the Company to have no government loan at all, rather than that Brunel should be enabled to finish the work.

Vignoles's plan, together with a model, was put on view at the company's offices in June, a week before a general meeting of the shareholders, at which Smith hoped to force a decision to get rid of Brunel once and for all. Few details of the plan have come down to us, and those chiefly through the criticisms of others, notably the younger Brunel, who considered that there were 'insurmountable objections to such a plan being carried into effect'. It appealed to Smith and his party because it would save money by dispensing with the 'shield', the ingenious invention by which Brunel had mainly overcome the problems of tunnelling through the water-logged soil of a river-bed. But Vignoles was only putting back the clock; and in essence his plan merely repeated one of the several unsuccessful schemes which had preceded Brunel's. Ironically enough, it was Vignoles's own grandfather, Dr Hutton, who as long ago as 1808 had declared these schemes impracticable, when they had been referred to him and William Jessop for criticism.

As a result of the general meeting, described by Brunel as 'excessively animated and virulent', the shareholders resolved to treat with any party willing to carry on or contract for the work. This meant that Vignoles was given permission to go ahead with his experiments. He had been growing more and more impatient for a decision, and he could hardly help being exultant. Smith claimed at the meeting that the Duke of Wellington had said he was ready to give his preference to the cheapest plan, provided that it had the approval of an expert, a point taken up in a press report of the meeting. Vignoles wrote triumphantly to Mary:

. . . the Duke has already publickly declared that . . . he knows no difference between Mr Brunel or Mr Vignoles: and added that if Mr Vignoles Plan could be effected for £100,000 less than Mr Brunels then Mr Vignoles was the Man for him: you will see this in the Morning Herald of the 1st July. . . . Indeed my friends here when joking call me 'His Grace's Man'!

His Grace was quick to deny the imputation, at any rate to Brunel, who wrote on 9 July: 'The Duke of Wellington sent for me and the object was respecting the Tunnel, that he had not said what the Gentlemen had said as being said by him.' Considering the circumstances, Brunel remained remarkably patient and unruffled, and gave orders that Vignoles should be given every facility at the tunnel.

There is no record of any report on the experiments, and for the rest of the year (1829) things hung fire. Despite the shareholders' decision, the Directors were still divided, and in any case all plans hinged on the question of a government loan. Vignoles left London for work in Ireland, and Brunel took a few months' holiday in France.

It was not until 26 February 1830 that Vignoles formally submitted his final proposals, which the shareholders considered at a meeting on 2 March. They seemed however reluctant to take the final plunge. When at last on 23 March the Board referred the plan and estimates to a sub-committee of directors for them to report on in the following year, Brunel finally lost patience and resigned. The effect was electric. The Board, realising at last the necessity for firm decision, invited a panel consisting of Tierney Clarke, James Walker and Peter Barlow to comment on Vignoles's proposals. Clarke and Walker were experienced engineers for both of whom Vignoles had worked. Barlow was a colleague of Dr Gregory at Woolwich. No three men could have had more reason to be fair to the engineer whose work they had to judge.

After two months' deliberations the panel concluded on 22 June that it would be a waste of time and money to attempt to complete the tunnel on Vignoles's plan. 'No other plan than that of Mr Brunel should be used to complete the Tunnel – and the directors should be instructed to apply to Parliament for a loan.'

Vignoles was in Liverpool when he received a copy of the arbiters' report. We have no record of what he thought of it; but as it came hard on the heels of his appointment as engineer to two projected railways in Lancashire, perhaps he did not worry unduly. The fact was that the Thames Tunnel had proved to be a diversion from the main stream of the career which he had begun to build for himself in railway engineering. During the last two years he had not lost touch with his Liverpool friends. At their request he had prepared plans for a road bridge over the Mersey at Runcorn; and his interest in the Lancashire railway scene had been vigorously renewed when he became involved, in the summer of 1829, in the famous locomotive engine trials on the L.&M.R. at Rainhill. Since then he had surveyed two railway lines and piloted their Bills through Parliament. It seems more than likely therefore that he was glad to be able to withdraw from an affair which had reflected credit on no one but Marc Brunel.

The arguments between the supporters of locomotive and stationary engines had come to a head with the publication in March 1829 of a report by James Walker and J. U. Rastrick which favoured the use of stationary engines, although they acknowledged that there might be improvements in locomotive engines in the future. The L.&M.R. Board was still divided on the issue; but the supporters of the locomotive were able to carry a resolution offering a prize of £500 to discover the most improved type of locomotive. This would not necessarily commit the Company to its use, but the locomotive party hoped that the trial would demonstrate the capabilities of locomotives as such – which in the event is what happened.

Vignoles shared Stephenson's opinions of the merits of the locomotive, and realised that any machine built by his firm would be a strong favourite for the prize. Yet there were other competitors, eager to beat 'Old Geordie' on his own ground. Vignoles had little mechanical knowledge, but he was a quick learner and always ready to take up a challenge. He threw in the support of his keen if inexperienced mind, and such money as he was able to raise, with two young engineers from London, John Braithwaite and John Ericsson. Braithwaite had a small engineering business in New Road, St Pancras; Ericsson was a Swedish army officer who was to have a great future as an inventor and engineer. Together they had recently produced the first practicable steam fire-engine, which had already made several successful appearances in the London streets. Barely three months before the date fixed for the trials they had decided to build a locomotive to enter the contest.

Five machines took part in the trials, but only two were to present any serious challenge to Stephenson's *Rocket*, the *Sans Pareil* of Timothy Hackworth, and the *Novelty* of Braithwaite and Ericsson.[2] In the eyes of many of the ten thousand or so spectators gathered at Rainhill on 6 October 1829, the *Novelty* was the favourite. Its boiler, water-tank and cylinders of burnished copper, and body-work decked out in royal blue, naturally appealed to a populace bred to admire the shining leather and brass-work of the gentry's carriages, in contrast with the plainer and more rugged appearance of the *Rocket*. According to the *Morning Herald* 'the velocity at which it moved surprised and amazed every beholder. . . . It actually made one giddy to look at it'. On a trial trip, drawing a train of wagons with 45 passengers, it averaged 22 miles per hour, with a maximum speed of 32 m.p.h. But when it came to the test it proved to be altogether too light for the work, and broke down. George Stephenson's often quoted comment: 'Eh mon, we needn't fear yon thing, her's got no goots' summed it up, though at the time the immediate cause of its discomfiture seems to have arisen from hasty workmanship.

'Vignoles doubtless hoped to see it run rings round the *Rocket*.' So writes L. T. C. Rolt of the *Novelty*, in his excellent and lively account of the Rainhill Trials, in his life of George and Robert Stephenson. Vignoles certainly applied himself to its support with characteristic enthusiasm, riding on the engine, timing each run, and compiling a series of notes from which Olinthus Vignoles prints extracts, but of which the original text has disappeared. Both the *Liverpool Albion* and the *Morning Herald* acknowledge Vignoles as an important source of information, and Olinthus Vignoles assumes that his father supplied material to the *Mechanic's Magazine*. This may perhaps explain the partiality of the

Press towards the 'London engine', as well as the excuses they offered for its breakdown. The phrase itself had echoes of the old Stephenson–Vignoles conflict and the rivalry between the engineers of north and south. In the end, however, after a week full of mishaps and adventures, only the *Rocket* fulfilled all the conditions laid down for the trials, and was declared the winner. Even Mr Robertson, the editor of the *Mechanic's Magazine*, did not dispute this decision, though he suggested that with improvements the *Novelty's* capacity for raising steam quickly would ensure its future success compared with the *Rocket*. The latter could only produce the same power by being larger and heavier, and carrying a higher chimney, factors which made for heavy wear and tear on the rails and instability of the machine.

There was something in this; but subsequent developments, particularly in the strength of permanent way, would show that in the long run weight would tell, and the *Rocket*, rather than the *Novelty*, became the ancestor of the modern steam locomotive.

However, the *Novelty* was not yet out of the running. Its owners refused an offer from the L.&M.R. Company to buy it, once it was repaired, preferring to retain it for their own experiments. Various modifications were made, and at a very successful trial on 17 December it reached a speed of nearly 40 miles per hour. In a letter to Braithwaite, Ericsson claimed that he could have done a mile a minute if he had dared trust the force pump at such a rate. The final result was that the L.&M.R. Company ordered two engines from its builders at £1000 each – to be ready by 1 July 1830 – in addition to the eight ordered from Stephenson.

Certain features of the *Novelty* are worth recording. It had a pair of vertical cylinders, whose pistons drove a double crank-shaft, and it carried fuel and water on the engine-frame, thereby qualifying as an early example of the tank engine. In the lithograph 'Novelty and Train', engraved by Robert Martin from a drawing by Vignoles, a basket of coke is shown standing ready amidships, while close examination reveals that the well-dressed engineer is in the act of lifting the lid of the coke-hopper on top of the boiler with one hand and adding a handful of coke with the other. The arrangement of boiler and flue was unique, the boiler being in two parts, one vertical, the other horizontal, running from end to end

The *Novelty* and a train of carriages. Lithograph by R. Martin, from a drawing by Charles Vignoles, 1829.

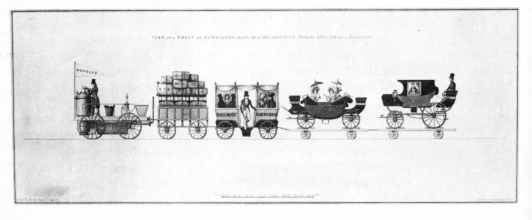

of the oak frame. Through the horizontal component the flue made a triple journey, finally emerging in a small chimney at the end opposite the furnace. One advantage of this system must have been that the men on the footplate never suffered from cold feet!

Speaking in 1870 as President of the Institution of Civil Engineers, Vignoles recalled how the *Novelty*, the 'beau ideal of the locomotive . . . if it did not command success, deserved it'. He assisted at the further trials, and for many years maintained a proprietary interest in it.

The success of Stephenson's *Rocket* at Rainhill set the seal on the public interest in the possibilities of the railway, aroused by the passage of the Liverpool & Manchester Railway Bill. Thus the coal owners of St Helens launched a scheme for a line from St Helens to the Mersey at Runcorn Gap, to break the monopoly of the coal carriers of the Sankey Canal. In addition the towns of Wigan and Warrington, north and south of the L.&M.R., saw the advantages of being linked with it. These branches, planned for local traffic, would ultimately become part of the main line from London to Carlisle; and there were a few men of vision in Liverpool who were already thinking of how they might travel by rail from their city to Birmingham or even to London. One way would be to extend the Warrington line southward, a route favoured by shareholders of the L.&M.R., since it would involve running over their line as far as Newton Junction; the other would be to follow a more direct route from Liverpool, bridging the Mersey at Runcorn.

Most of the promoters of these lines were Liverpool men, some already shareholders of the L.&M.R. When it came to surveying the new lines and getting them through Parliament, they did not have to look far for an engineer. The ones they knew best were Stephenson and Vignoles, each of whom had their supporters. Stephenson's son Robert, who had recently returned from America, was also in the running. In the end, while Robert Stephenson became engineer of the Warrington line, those of Wigan and St Helens fell to Vignoles; and Vignoles was chosen to survey the direct line from Liverpool to the south, via Runcorn. This was to be known as the Liverpool & Chorlton Railway. (See map, p. 27.)

Great public excitement was aroused by these schemes, and the sense of urgency was shared by the engineers. No sooner was Vignoles appointed than he was hard at work on the preliminary surveys, having, in the manner of a general mobilising his forces, called in his assistants from London and Ireland to join him on the ground. The Parliamentary plans had to be ready for deposit in Parliament, and with the Clerks of the Peace for the county, by the end of November. In a letter on 4 December Vignoles reported to his wife that he and his assistants had not been to bed for three nights on end:

The hurry anxiety and business was five times what it was with the plans for the L&MR, and when all was finished everybody was worn out. I have however accomplished my Task but it has left me full of nervousness and has reduced me to a Skeleton and what is worse I see no End to it. The public estimations and enthusiasm for the Railways and Locomotive Machines is daily augmenting and I find my Services and Opinions are

Table 1. *Estimates for a double line of railway*
(written on the flyleaf of Vignoles's diary for 1830)

General Estimate for a <u>Double</u> Line of Railway

			per Mile	
Blocks and Sleepers			700	
Stoning, Metalling & Boxing			650	
Rails	1200			
Pedestals	400	Iron	1750	
Keys	100			
Bolts	50			
Laying Rails			<u>400</u>	
			£3500	Railway laid complete
Land from £800 to £1200		say	1000	
Outer Fences & Drains			350	
Inner Drains & Soughs			<u>150</u>	
			£5000	Exclusive of Fencing

Common Fencing, Culverts.
Gates, Lodges & ordinary
Masonry, etc. etc. say   1000
                                                       £6000

If the Excavations are heavy the Masonry etc. is increased.

St. Helens Raily. Excavns. per Mile	say	£2400	
Wigan Branch Rly. Do.		£2000	Average say
Preston Rly. Do.		£7000	to
Liv & Man Rly. Do.		£8000	£5000
St. Helens Masonry		£1500	
Wigan Br. Rly.		£2000	Average
Preston Rly.		£2500	say to
Liv. & Manch. Rly.		£3500	£3000

Double lines of Railway

The Average Cost of the St. Helens & of the Wigan Railway
exclusive of Land & Iron has been as nearly as possible

	£6000 p. mile	
Average Cost of Iron	1750	
Cost of Land & Drainage for St. H.	£1000	1150 p. mile
Wigan Ry.	1300	
The Act of Parliament	<u>850</u>	
	£9750	

[The punctuation, spelling and layout are Vignoles's own.]

in great requisition.

He could not resist adding, 'Mr Stephenson is losing Ground as fast and perhaps faster than I am gaining it', and with a characteristic touch of arrogance: 'I now want nothing but the Tools in order to have half the Railways in the Kingdom under my Control.' Finally he praises the work done by his principal assistants, Forth, Terry and Collister, in proper military terms: 'all my assistants have done their duty'.

At the beginning of 1830 Vignoles resumed the diary which had been broken off in 1825 during his survey of the L.&M.R. For over 30 years he was to maintain this daily record of his work, with occasional entries concerning his private affairs. The wide variety of his activities at this time, recorded in his journal, is remarkable. He was planning, if not half the railways in the kingdom, at least a dozen different lines, of which only the St Helens and Wigan lines were eventually constructed. He was rapidly familiarising himself with the techniques of surveying, costing and laying out railways. Thus on the flyleaf of the 1830 volume is a tabulated estimate for a double line of railway, broken down into its various items (see Table 1).

On the mechanical side, in addition to the *Novelty* tests, he conducted trials with the Irish engineer Alexander Nimmo of a new low-pressure boiler 'on the exhausting principle' developed by Braithwaite and Ericsson. An account of these trials appeared in the *Mechanic's Magazine* for 5 June 1830, and on 1 May 1830 this journal also published tables drawn up by Vignoles showing the power developed by locomotive engines at different speeds and gradients. With John Ericsson he evolved a method by which a locomotive would ascend an inclined plane by means of a pair of rollers exerting pressure on a centre rail. A patent for this

'Skew' bridge on the St Helens & Runcorn Gap Railway, crossing the Liverpool & Manchester Railway, 1832. Aquatint by S. G. Hughes. The artist has used his imagination in depicting the locomotive on the bridge, vaguely resembling the *William the Fourth*.

was filed on 7 September 1830, and the inventors hoped to apply the principle to the propulsion of canal-boats in tunnels. (Vignoles went so far as to offer the idea to Captain Bradshaw.)

These ventures into mechanics were an essential adjunct to his civil engineering work. It was particularly important that railway builders should understand the capabilities of locomotives in tackling curves and gradients. Vignoles was one of the first engineers to grasp the importance of 'banking' on curves, and though he incorporated the principle of inclined planes, with stationary engines, into his surveys of the L.&M.R. and the St Helens railway, he was soon to appreciate the possibility of locomotives climbing quite steep gradients unaided. Thus on 4 September the diary records experiments with Bury's *Liverpool* engine, which after beating Stephenson's *Dart* and *North Star* climbed the Sutton and Whiston inclined planes drawing a load of 20 tons weight, from which Vignoles calculated an average speed of 8 m.p.h. on an incline of 1 in 96. The design of permanent way was also much in his mind. In November of this year he was to meet Robert Stevens, the American engineer who introduced the flat-bottomed rail into the U.S.A., a meeting which probably marked the beginning of Vignoles's interest in the rail section which was later to bear his name.

The Parliamentary bills for the St Helens & Runcorn Gap Railway, and for the Wigan Branch Railway, received the Royal Assent on 29 May 1830. They had not been passed without opposition, and there are several references in Vignoles's correspondence with his engineering friends to 'enemies' and 'intrigues' concerning his appointment as engineer to these lines. However he was unanimously elected Chief Engineer at a meeting of each company, on 15 and 16 June, at salaries of £650 and £500

Seal of Wigan Branch Railway Company, 1830. Designed by Charles Vignoles, showing the locomotive *William the Fourth*.

per annum. As in each case he was to find and pay his own assistants and pay his own expenses, these were hardly princely figures. He was expecting, however, to receive £2300 for his preliminary survey and Parliamentary work, most of which he had already spent on expenses.

The St Helens line was a small affair, initially a single track with passing places, its total length being about eight and a half miles. Branches to various collieries en route and to the St Helens glass works, and the junction with the L.&M.R., made up another seven miles of track. There were no tunnels or viaducts, but swing bridges at the points where the line crossed the Sankey Canal. This avoided the necessity of raising the level of the line at the crossing. At the point where it crossed the L.&M.R., however, Vignoles designed a skew cast-iron bridge, with an inclined plane on either side. Stationary engines with ropes took over from the locomotives at this point, as well as at the inclined plane where the line ran down to the new Widnes Dock. The negotiations for the skew bridge, and for junctions between the two lines, gave Vignoles the satisfaction of dealing with George Stephenson as one chief engineer with another.

By the end of August the first sods had been cut. Messrs Pritchard & Hoof (freed from any obligations in connection with the Thames Tunnel) contracted to build the Wigan branch, and Messrs Thornton the St Helens line, while negotiations for the supply of rails, chairs, cast-iron bridges and bricks were being completed with various foundries and factories in Lancashire and Yorkshire and as far afield as Bristol. Vignoles also concerned himself with such details as the design of the company seals and share-headings. Meanwhile the cotton-spinners of Preston joined with the promoters of the Wigan branch in a plan to extend the latter northward as far as Preston, providing them with a rail link with Manchester and Liverpool. It was only logical that Vignoles should become the engineer for this extension, which later amalgamated with the Wigan branch to become the North Union Railway.

On 15 September 1830 the Liverpool & Manchester Railway was officially opened, amid scenes of great excitement. Vignoles was one of the many official guests who travelled on eight special trains from Liverpool to Manchester and back. Many accounts have been written of this memorable day, graced by the presence of the Duke of Wellington, and marred by the fatal accident at Parkside Station to William Huskisson, Liverpool's M.P. Vignoles once again supplied notes for the *Mechanic's Magazine*, while he recorded in his diary, with what Olinthus Vignoles describes as the 'brevity characteristic of the soldier in the presence of some great calamity':

Opening of the Liverpool and Manchester Railway. Procession went from Wavertree Lane to Parkside Bridge being 14¾ miles an hour. Then Mr Huskisson was killed by the 'Rocket' Engine passing over him.

The procession went on to Manchester. In returning the procession occupied 5 hours – viz. 2½ hours from Manchester to Parkside Bridge with 3 Engines and 24 Carriages and 2½ hours from Parkside to the Tunnel Head with 5 Engines and the same Load. N.B. All the Gentlemen had to walk up the Inclined Plane.[3]

On 24 September we find the equally laconic entry: 'Returned to Liverpool for the funeral of Mr Huskisson.'

Unfortunately for their makers, the *Novelty's* successors were not completed in time for the L.&M.R. opening, and the two engines, *William the Fourth* and *Queen Adelaide*, proved unable to compete with Stephenson's locomotives. The forced draught from below was replaced by a fan wheel turning in a copper vase at the top of the boiler, which Ericsson himself acknowledged to be 'very classical' but 'miserably inefficient'. Eventually the two London engineers accepted defeat, and turned their attentions in other directions. But the experience of working with them inspired Vignoles to issue his own specifications for a locomotive, which was circulated for tender by the St Helens Company early in 1832. The cheapest quotation of £455 came from the Horseley Coal and Iron Company of Tipton, and the first three engines were ordered from them.[4]

For Vignoles the opening of the L.&.M.R. was an interlude in an intensive programme of work, during which he was constantly on the move. For example, on 1 October he examined the whole course of the line from Wigan to Preston on horseback, and made final decisions on its course. On the next day he was out inspecting the St Helens line. So it went on to the end of November, by which time the Parliamentary plans for the Wigan & Preston line were completed. Vignoles and the company solicitors posted by night to deposit them at the county offices. (The diary notes: 'Chaise from Wigan to Preston £1.2.6. Gates 4/6. Drivers 10/-. Ostler 1/-'.) The Bill and other matters took him frequently to London, a coach journey of nearly 24 hours. At the end of January 1831 he was journeying through snow to hunt up owners of Lancashire land who were resident in Berkshire and Sussex. Two weeks after returning to Liverpool he travelled south again. This time he went by rail to Manchester, then by mail-coach to Northampton and on through the night to Barnet, from where he posted in the early hours to London. This was probably the first time Vignoles made part of the journey by rail. The fare was 5s.0d. A couple of days later he returned to Liverpool by coach, worked hard for five days, then took the mail-coach once more for London. On 9 March he was back in Liverpool, and on 13 March he was again in London, for the Parliamentary proceedings which were to open the next day.

The *William the Fourth* locomotive, made by Braithwaite & Ericsson, 1830.

This was a pressure of work and travel which called for physical and mental stamina of no mean order, and it was to be the pattern for long periods of Vignoles's working life; and if his handwriting is anything to go by he may well have relieved the tedium of long journeys by using the time to write up his diary.

On 21 April he wrote to his wife, informing her that the Wigan & Preston Bill was about to receive the Royal Assent 'in spite of all exertions of our adversaries to the contrary. We are thus the *only Railway Bill* which has passed this Session.' As it was the session of the Reform Bill, perhaps Parliament could be excused for not having given much attention to railways. Characteristically, Vignoles does not mention the cause of reform, nor the excitement and uproar it was arousing in Parliament and the streets of London.

Vignoles's Liverpool & Chorlton Railway Bill was also considered at this session, but was thrown out when the King prorogued Parliament on 22 April. Vignoles continued to work on the plans for the very large bridge over the Mersey at Runcorn, in which he had some assistance from George Rennie. But at the end of July the promoters gave way in favour of those who preferred the alternative line via Warrington, which was to become the Grand Junction Railway to Crewe Hall and Birmingham, and Vignoles turned over all his drawings and papers to the Grand Junction, whose engineer was his former colleague on the L.&M.R., Joseph Locke.

During 1831 work on all three of the lines proceeded steadily. The St Helens & Runcorn Gap Railway was the first to have a portion of its line opened, though the Wigan Branch Railway was the first to be completed. The latter seems to have offered few problems in its five miles of length, while progress on the St Helens line was so good that by the end of the year Vignoles was able to arrange for the first train of coal to be brought from the Broad Oak Colliery along the completed section of the line to Sutton on the L.&M.R. and thence to Newton Junction and by way of Robert Stephenson's new line to Warrington. Vignoles had hoped to use the *Novelty* for this ceremony, but unfortunately, during experiments, it showed a tendency to leave the rails. In its place the *North Star* was used, on hire from the L.&M.R., with wagons supplied by the Warrington Company. On 2 January, the first train of coal and passengers proceeded from Broad Oak to Warrington. Later in the day, a second train ran through to Liverpool, where a dinner for the directors and shareholders marked the completion of the first stretch of line entirely carried out by Vignoles.

5 The Dublin & Kingstown Railway and the Irish Railway Commission, 1832–1839

By the beginning of 1832 Vignoles could pride himself on being established as a railway engineer of consequence. He had two railways under construction, another in the surveying stage; the plans of several more were in mind. And he was looking further afield than England. He had kept in touch with his first employer in America, Major W. Hasell Wilson now engineer to the Columbia & Philadelphia Railway; American engineers were showing great interest in railway development; and Francis Ogden, the American steamship promoter and U.S. Consul in Liverpool, had enlisted Vignoles's help for visitors from across the Atlantic. There is a reference in the latter's diary as early as 1830 to a consultation with Braithwaite about a possible railway from Paris to St Cloud. Now he was invited to meet representatives of the 'Ponts et Chaussées' (Public Works Department) in Paris, to discuss a plan for a railway from Paris to Dieppe. At once he saw this as part of a London–Paris link, in conjunction with the Brighton line in which he was still interested.

Back in England he began work on a report to the French Government. Long before it was completed, a new field of activity opened up for him, in Ireland.

When Vignoles had left London for Ireland in August 1829 (see Chapter 4 above), he had been commissioned by the Earl of Portsmouth's trustees to survey the Earl's Irish estates, and to advise on possible improvements to the Enniscorthy Canal, in County Wexford. These lay not far from his birthplace at Woodbrook, where his grandfather William Blacker put a room at his disposal in the house where he had been born. It was the first time Vignoles had set foot in Ireland since he had left it as a baby for the West Indies, but he liked to call himself an Irishman when it suited him to do so. He wrongly believed that he was related to the Irish family whose most important representative at this time was Dr Charles Vignoles, Dean of Ossory and Chaplain to Dublin Castle. These contacts, and others, were to lead to his appointment as engineer to Ireland's first railway – the Dublin & Kingstown.

It had originally been intended that the city of Dublin should be linked by a canal with Kingstown (or Dunleary) Harbour (begun by the elder John Rennie in 1816), but the project was judged to be too costly, and ultimately, in 1831, an Act for the construction of a railway was obtained, and Alexander Nimmo was appointed Chief Engineer. A fair proportion of the capital of £150 000 was subscribed before the Bill reached Parliament. Nimmo died in January 1832, whereupon first Telford and then George Stephenson were invited to report on the plans. The former declined because of ill-health, and although Stephenson reported favourably the recently-constituted Irish Board of Public Works (who were to advance £75 000 to the Company) suggested that their own engineer, Killaly, should make a further independent report. On 6 April Killaly also died, after a brief illness. Before his death, approaches had already been made to Vignoles, who was well known to the Board because of his work on the Enniscorthy Canal. On the 7th he was summoned to Dublin to survey the line and report on Nimmo's plans. After a hard week's work studying the line with his most senior assistant, Charles Forth, and the

Charles Blacker Vignoles, aged 38 years. Engraved by R. Roffe in 1835, for the *Mechanic's Magazine*, from a miniature painted *c.* 1831, artist unknown.

Irish contractor William Dargan, Vignoles delivered his report, and a revision of Nimmo's estimates. As a result he was invited to prepare his own detailed proposals for building the railway. The general line, as laid down by Nimmo, had been fixed by the Bill. Having a total length of five and a half miles, it started as an elevated line in the city, and ran down at a gentle gradient onto flat meadows and then along the sea-shore of Dublin Bay. For most of the length the land to be acquired was restricted by the Act to a width of a hundred feet. The main engineering works were the high-level passage through the Dublin streets after leaving the terminus at Westland Row; the crossing of the River Dodder, which was liable to severe flooding; and the construction of a stable embankment to support and protect the line where it skirted the waters of the bay. A particular problem was offered to the line by the private grounds of the two principal landowners on the route, who had been specially exempted from the compulsory sale of their property.

Immediately after presenting his report, Vignoles returned to Liverpool, taking Dargan with him to view the various Lancashire railway works, and set his assistants to work on the Irish railway plans.

A new travel dimension – the crossing of the Irish Sea – now entered his work. Forth and another assistant were put in charge of the detailed work in Dublin, while at Liverpool a team of half a dozen assistants toiled at the drawings and plans in addition to the work they already had in hand on the English lines. Vignoles went to and fro between Dublin and Liverpool by the auxiliary steam packet, usually by night, often in stormy weather, plunging straight into work on arrival at either end; conferring with directors, riding and walking over the ground, supervising, revising and correcting; the Dublin and Lancashire works proceeding simultaneously. In addition the Paris report was completed and sent off in June, while in Ireland he was drawn into consulting work left unfinished by the death of the late government engineer. From the beginning of April to the end of July he crossed the Irish Sea 19 times, and the total of single crossings reached 35 by the end of the year.

On 18 June he presented his new plans to the Directors, and on the 19th to the Board of Public Works, supported by James Pim, Jnr., the Company Treasurer. One of a family of Dublin bankers, and a keen protagonist of railways, Pim was to form a lifelong friendship with Vignoles. The chairman of the three-man Board of Public Works was Colonel John Fox Burgoyne, R.E., son of General John Burgoyne, the famous 'Gentleman Johnny', unfortunately better remembered for his surrender to the American rebel army at Saratoga than for the unlikely fact that he was a dramatist as well as a soldier. The younger John Burgoyne was a military engineer of some capacity, at present seconded to government service but destined to reach the highest military rank.

Although the statement by Olinthus Vignoles that Burgoyne had been with Charles Vignoles 'in some of his military campaigns' is manifestly untrue, there is no doubt that the two men were drawn together by professional interests. For many years their friendship flourished on this basis, surviving many disagreements and the expression of forthright opinions. It was probably Burgoyne who decided that Vignoles should be deputed to tackle Lord Cloncurry and Sir Harcourt Lees, the two aristocratic landowners whose objections to the railway passing through their estates were threatening to ruin the whole undertaking.

It was a pleasing characteristic of nineteenth-century engineers that they believed that their work should have artistic as well as engineering

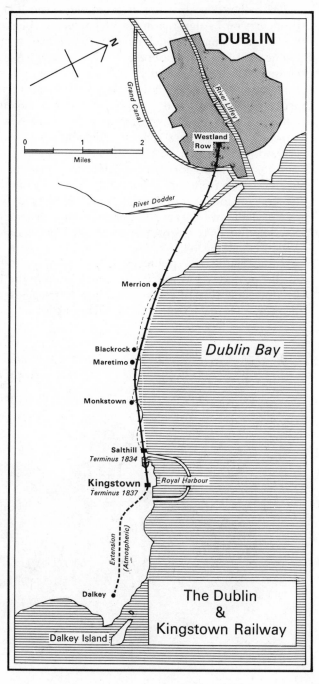

merit. Bridges, viaducts, tunnels, embankments were constructed with an eye to the landscape, and decorated according to the architectural taste of the period. Vignoles was no exception, and it is significant that he often used the term 'works of art' as a general description of his bridges and viaducts. Faced with the opposition of Lord Cloncurry and Sir Harcourt Lees, he endeavoured to persuade them that the railway could embellish rather than disfigure their grounds. He called on Lord and Lady Cloncurry on 1 August, armed with models and plans. His first intention had been to tunnel through the sloping park which ran down to the sea, but he was obliged to adopt the cheaper alternative of a partially covered cutting, which he assured his Lordship would effectively conceal the trains from his home; a latticed bridge of handsome design would give access from the park to a 'maritime promenade' beyond the railway, with bathing lodges in the classical style, a pier and a small harbour; and the entrance to the cutting, as befitted the approach to so important a property, would be flanked by a pair of lodges or pavilions of italianate design. The noble lord was not at first convinced. He maintained that the benefits likely to arise from the undertaking did not justify the sacrifice he was being asked to make. In the end, however, he gave way, in consideration of a payment of £3000 in addition to the embellishments to his property. Sir Harcourt Lees was pleased to accept £7500 and a bridge, but got no towers or bathing lodges.

Although the conception belongs to Vignoles, the architectural design of the 'pavilions' has usually been attributed to the Dublin architect J. S. Mulvaney. Recent research has however suggested that they may be the work of William Cole, Jnr., a young English architect whom Vignoles

Lord Cloncurry's pavilions at Maretimo, Blackrock, on the Dublin & Kingstown Railway, 1834. One of the *Thirteen Views on the Dublin & Kingstown Railway*, drawn by A. Nichol, engraved by S. G. Hughes. The artist has given his own idea of the locomotives and carriages, and placed them on the wrong lines.

engaged as an assistant. Evidence for this is to be found in Vignoles's diary, which states that Cole accompanied Vignoles to Dublin in May 1832, and spent a week examining the route through Lord Cloncurry's grounds, seeing how to 'conciliate with proper designs'. Vignoles also writes on 17 May that five draftsmen were working in his Liverpool office, on the plans and a model of the 'Cloncurry passage', and states specifically that Cole was in charge of the architectural designs. Nowhere in the diary is there any mention of Mulvaney. Further evidence of Cole's involvement, which is entirely consistent with Vignoles's practice of picking the brains of experts wherever possible, is to be found in a letter from Cole to Vignoles dated 30 May 1832.[1]

All this cost money, a commodity of which the Company was short, and it seems that the directors were not altogether satisfied with Vignoles's estimates, for they called in J. U. Rastrick to give his advice. As a result of his recommendations, Vignoles was obliged in December to make a thorough revision of the specifications drawn up for the contractors.[2] This done, on 26 January 1833 the Directors awarded the contract to build the line to William Dargan, for the sum of £83 000, the work to be completed on or before 1 June 1834. As from the New Year Vignoles was to be engaged on an annual salary, he submitted his account to date: £3500, of which £500 was for fees, the rest being his outgoings for assistants, tradesmen, models and other expenses.

During all this time, affairs in England had continued to press. On 3 September 1832 the Wigan Branch Railway was opened, Stephenson's *Rocket* being hired to draw the first train. (Vignoles recorded a 'tip' of £1 given by him to the engine-driver.) Progress on the St Helens line had

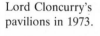

Lord Cloncurry's pavilions in 1973.

suffered a set-back at the end of May, when the dam at the Widnes Dock gave way, and Vignoles had to work all night with a party of workmen to prevent the Sankey Canal from being flooded out. Further delays arose through slow work by the contractors, but at the beginning of February 1833 it was possible for the first locomotive to travel the whole way from St Helens to Runcorn Gap. The line was officially opened on 21 February, without ceremony, as befitted a railway dedicated mainly to the carriage of goods.

Meanwhile the Dublin & Kingstown Company were applying for a Parliamentary Bill to authorise the extension of their line as far as the new mail-packet wharf at Kingstown, involving the construction of an embankment across a corner of the old harbour; and to continue the line

Table 2. *Copy of flyleaf of Vignoles's Diary for 1833*

Charles Vignoles, Civil Engineer
Harrington Chambers, Sth. John St,
Liverpool
No. 3 Westland Row, Dublin
No. 4 Trafalgar Sq, London

Dublin & Kingstown Railway.   Capital £200,000.

Traffic on the Rock Road out of Dublin for one Year 1831–1832

Private Coaches	36,287 avge.	2 persons each	72,574
Hackney Coaches	7,272	4	29,088
Private Cars*	133,537	4	534,148
Public Cars*	186,108	4	744,432
Gigs	24,175	2	48,350
Saddle Horses	46,164	1	92,328
			1,520,920

Besides a vast number of persons on foot.

÷ Days 365   4,167
in round numbers   4,000

4000 at 6d = £100 p.d = £36,500 p.a.
Expenses & interest on loan say   £11,500 p.a
£25,000

= 5% on Capital of £500,000

*[i.e. Irish 'jaunting cars'.]
[The punctuation, spelling and layout are Vignoles's own.]

as far as Dalkey, where there were valuable stone quarries. The hearings of the Parliamentary committee were marked by a clash of personalities which took place between Vignoles and the Chairman, Daniel O'Connell, M.P. The dispute arose over the estimates made by Thomas Woodhouse, the resident engineer, which had been grossed up to round figures by the addition of a sum for contingencies. The famous Irish nationalist maintained that such a sum could only be an arbitrary figure and criticised Vignoles for having sworn that the final figure was a true estimate. Vignoles protested hotly at what he claimed was an unwarrantable attack on his integrity as a gentleman, and though O'Connell stuck to his point, he was obliged to assure Vignoles that his remarks carried no imputation of perjury. The Bill was in fact thrown out for navigational considerations, mainly on evidence given by Sir John Rennie, at that time consulting engineer to the harbour authorities.

In the same session of Parliament the Bill for the London & Birmingham Railway, surveyed by Robert Stephenson, was finally passed, after having been rejected by the House of Lords the previous year. Typically, Vignoles applied for the post of Chief Engineer to the line (possibly encouraged by the anti-Stephenson lobby in Liverpool), only to be informed on 29 June that Robert Stephenson had been appointed. Yet he had enough work on hand to satisfy the most energetic of men. At this very moment he was engaged on a further examination of the London to Brighton route, and he had managed to interest his Irish friends, particularly the D.&K.R. solicitor, Pierce Mahony, in his dream of a Paris to London railway, via Dieppe and Brighton. With their help, Vignoles engaged the support of the Duke of Richmond, one of the principal landowners of the counties through which the London and Brighton line was to run, as well as the financier Moses Montefiore. It was decided to retain William Cubitt as consulting engineer, and to despatch Vignoles to Paris to sound out French opinion. Armed with a quantity of plans, reports and letters of introduction, Vignoles left London on 12 August. In Paris he obtained a series of interviews with Comte le Grand, Director-General of the Ponts et Chaussées, M. Adolphe Thiers, Minister of Public Works, and other members of the French Government, and was received at St Cloud by the King, Louis-Philippe. In return for an assurance by Vignoles that the London and Brighton line was certain to go ahead, Thiers promised to recommend to the French Government that they should back the Paris–Dieppe project and co-operate with the British in raising the necessary capital. He also accepted an invitation to a tour of English engineering works, which Vignoles hoped would finally convince him of the merits of rail transport.

During the two weeks of his visit Vignoles found time to see the sights of Paris, Versailles and St Cloud with his 15-year-old daughter, whom he had placed with a French family to complete her education. He also wrote long letters to Mahony reporting progress;[3] and with his usual capacity for interesting himself in a diversity of projects he drafted proposals for improvements in the Paris water supply.

On his return to England he spent a busy five days reporting the results of his visit to the London supporters, while making preparations to receive M. Thiers. The latter arrived in England on 6 September, accompanied by the Comte le Grand and other officials. Thiers had insisted on bringing over with him his own cumbersome double-decker coach, which required six post-horses to pull it. In this vehicle, 'quite *à la Louis quatorze*', as Vignoles described it, Thiers and two of his staff (le Grand was content with visiting London) were conducted by Vignoles on a tour covering five days and nights, of which three nights were spent on the road. Highlights of the journey were Telford's Menai suspension bridge, trips on the Liverpool & Manchester Railway and on the St Helens line, Wedgwood's pottery factory, and the blast furnaces at Wolverhampton, which Vignoles, with his eye for artistic effect, was careful to show his guests at night. At Liverpool they dined with the Mayor, and as a change from engineering works the itinerary took in Conway and Warwick castles. Three more days were spent in London, inspecting water-works, the East India Company's warehouses and the Custom House, and Vignoles procured for his visitors copies of Acts of Parliament, railway prospectuses and other relevant documents.

The outcome was a disappointing end to six weeks' hard work. On parting with Vignoles, Thiers, with true Gallic candour, declared himself unable to see the advantages of rail transport, though infinitely obliged to Vignoles for all he had shown him; he did not think railways were suited to France. This opinion he expressed in a number of speeches on his return to that country, as a result of which the possibility of railway communication between France and England was pigeon-holed for a decade. On the other hand the plan for a London to Brighton line remained very much alive, for Montefiore and his City friends agreed to press on with the scheme with a view to obtaining a Bill in the next Parliamentary session, although John Rennie and others were known to be working on rival plans.

There was in fact little time for lamenting. In the south, the plans and sections of the Brighton line had to be completed for Parliament. In Lancashire the contractors' accounts for the Wigan Branch Railway and the St Helens railway were still to be checked, and various teething troubles sorted out. On the Wigan & Preston Railway detailed surveying and the designing of bridges, viaducts and other works were in full swing, while another Parliamentary Bill was being drafted to amalgamate it with the Wigan branch. And on top of all this, Vignoles had accepted the post of consultant to a new London & Windsor Railway, to report on the plans prepared for Parliament by their engineer, and recommend improvements.

Meanwhile in Dublin, in his office at 3 Westland Row, near the site of the railway terminus, a multitude of items was awaiting his approval: fences, gates, designs for the terminus buildings, for 'lodges' and 'boxes' for crossing-keepers and signalmen, for engine-sheds and water-tanks. Vignoles personally supervised the contract for the supply of rails. These, following the example of the L.&M.R., were to be laid in chairs mounted on stone blocks, with felt pads between them, the rails being tied together

with sleepers at intervals of 15 ft.[4] A Liverpool firm undertook to supply
platelayers and to maintain the track for two years. In a report submitted
to the Directors before he set out for France Vignoles had recommended
that at least eight engines should be ordered, built to a standard pattern,
with a plentiful supply of spare parts, so that repairs could be made when
required without upsetting the railway time-table. On 10 October he returned to the subject of engines, deploring the fact that the locomotive
sub-committee had been empowered to order only four engines, which
he claimed would endanger the efficient running of the line. The Company gave way to the extent of ordering six. The Horseley Company,
among other firms which tendered, was unable to promise delivery by
the date stipulated. In the end the order was equally divided between
Forrester's of Liverpool and Sharp, Roberts & Co. of Manchester. Both
firms made different modifications to Vignoles's specifications, thereby
sacrificing the principle of standardisation which he had recommended.

Severe autumn gales gave Vignoles and Dargan plenty to think about.
They were particularly anxious about that part of the line between Merrion and Blackrock, where it ran on a bank outside the low-water mark,
leaving a stretch of sheltered water between the line and the shore, ultimately to be reclaimed. They decided to raise the level of the line by six
inches, and on the advice of Colonel Burgoyne the embankment was
finished off with a convex parabolic curve. Pitched channels took the
place of covered drains to carry off surplus water from the line. After a
heavy gale at the beginning of November, Vignoles reported 'with infi-
nite satisfaction as an engineer and with high interest as an amateur
geologist' that the sea was consolidating rather than washing away the
material of the bank.

The company secretary, T. F. Bergin, who was himself something of
an engineer, had perfected one of the first types of spring 'buffing ap-

View of the
Dublin &
Kingstown
Railway between
Merrion and
Blackrock, 1834.
One of the
Thirteen Views
. . . , drawn by
A. Nichol,
engraved by J.
Harris. The view
shows the
embankment
enclosing a
stretch of water,
since reclaimed.

paratus' (i.e. buffer), and this was first used on D.&K.R. rolling-stock. He also collaborated with Vignoles in the design of passenger carriages, which were to be of three classes. By setting the wheels within the body of the carriage, beneath the seats, Vignoles contrived lower and broader steps than those on the L.&M.R., which he considered to be extremely awkward for 'invalids, elderly persons, females and children'. An unusual and superior feature of the third-class carriages was that they were to be roofed though open at the sides.

At the beginning of November, Dargan gave notice that delays in obtaining possession of the ground would make it impossible to meet the deadline of 1 June 1834, without much extra expense. Six months later a section of the line was still obstructed by the Salthill cliffs. However the obstruction provided an opportunity for the Chairman of the Board of Public Works and the Chief Engineer to enjoy a little fun together. Under Colonel Burgoyne's directions a train of powder was laid in a gallery excavated in the sandstone cliffs, and what Vignoles described as a 'very effective' explosion brought down about 200 cubic yards of cliff, at a cost he reckoned as slightly more than 1d a yard.

There had also been delay in the supply of materials. As late as 22 May no locomotive had been delivered by Sharp, Roberts & Co. When Vignoles called there he was annoyed to find no men at work (it was Whit Monday and they were all at Manchester Races), and 'to my regret locomotive engines were not forward . . . I mentioned to Mr Roberts that he shd. hasten them.'[5] Forrester's engines were more advanced. On 25 June Bergin and Vignoles tried out the first one on a stretch of the L.&M.R. line. 'With great care' a speed of 28 to 30 m.p.h. was reached on the level. 'The regular beating of the machine was delightful.'

Although it was still impossible to fix a date for the opening of the line, Vignoles advised the Directors that it should be a public affair, and outlined to them 'the manner in which a very handsome Entertainment might be given at moderate expense'. The Directors seem to have thought this a good idea; moreover they decided to go one better than the L.&M.R. by inviting the Princess Victoria, heiress to the Throne, and her mother, the Duchess of Kent, to grace the ceremony with their presence. Vignoles's connection with the late Duke gave him the *entrée* to the family. On 12 July he and Mahony waited on the Duchess and Princess, 'when I gave a description *viva voce* to the Princess of the Railway System in general and the Dublin & Kingstown Railway in particular and expatiated on the probable further extension of the benefits to Ireland.' The Princess was pleased to allude to the connection of Vignoles and his father with the Duke and the Duke's regiment, but no positive answer was given as to the royal party opening the railway 'on account of political embarrassments'. The Government had just fallen, over the question of Irish Church Reform and its attempt to bring in a coercion Bill to impose law and order on that unruly country.

As the summer of 1834 wore on, and the completion date seemed as far off as ever, Vignoles urged the contractors to greater efforts, while himself struggling with a thousand and one details, such as station gas

lamps, the fencing, clagging and coping of 'landing places' (i.e. platforms), the layout of terraces and walks in the Salthill grounds, where
there was to be an observatory and a camera obscura, and the police establishment 'ordered to be regulated under my Directions and drilled for
Service'. And, in addition to all this, at the back of his mind lay the private anxiety of his wife's ill-health.

Ever since Isabella's death and the illness which followed the birth of
Olinthus in 1829, Mary had suffered from recurring bouts of indisposition
and mental depression. In 1831 she had moved to join Charles in Liverpool, leaving the older children at a small boarding school in the Isle of
Man, but relations between them had only worsened. Their incompatibility was all too apparent. Mary was too cautious, too pessimistic, to give
Charles the encouraging support he needed. She resented deeply his continual absences from her side, necessitated by the irregularity of his professional life. Charles, heavily involved in the compulsive pressure of his
work and his ambitions for the future, was angered by her criticism and
fits of jealousy, which threatened to become an obsession. References
to medical consultations early in 1833 indicate that Charles was prepared
for his wife's possible mental breakdown. The blow fell in April that year,
recorded by the brief diary entry: 'In consequence of the extreme illness
of one of my family I was obliged to place the person with Mr Squires
under restraint.' Now, in September 1834, he was informed that there
was no hope for her recovery. An additional subject for worry was the
future of his five children, dependent, since Mary's illness, on the care
of friends, housekeepers and the proprietors of boarding schools, and on
such occasional attention as he was able to give them.

To relieve the strain of so much work and personal anxiety, he took
five days' holiday at the end of the month – an unheard-of event for him
– in the shape of a carriage tour through the Wicklow hills, with Camilla,
young Charles and the Liverpool bookseller James Mawdslay, one of his
oldest friends. Even then he worked for two or three hours each evening
on reports.

For most of October he was dealing with business in England. Back
in Dublin on 1 November he made his first trip on the railway, with a
locomotive and a train of carriages. Separate trials were made of the en-

Illustration from
Kirkwood's
*Dublin &
Kingstown
Railway
Companion*,
1834.

gines. Vignoles recorded:

I went one quarter of a mile with the *Hibernia* and her Tender only attached, made by Sharp Roberts & Co. of Manchester, at the rate of *Sixty Miles an Hour* !!!!! Mr Bergin went at the rate of 48 *miles an hour* with a carriage holding nearly 40 passengers, and for ¾ of a mile. Owing to bad Coke the *Vauxhall* made by George Forrester & Co. of Liverpool did not go more than 36 miles an hour for the same distance.

Less than a week later an unforeseen disaster struck the railway. After 24 hours of heavy rain, the River Dodder burst its banks, flooded the countryside and carried away two wooden bridges and the woooden centring of a stone bridge under construction, a short distance above the railway bridge. The mass of timber swept down upon the latter, jammed under the arches and raised them bodily from their foundations, which were also damaged by the scouring action of the water sweeping under the bridge. At the same time the flood poured over the track at a number of points. It happened that William Cubitt had just arrived in Dublin for consultation on Kingstown Harbour. Vignoles called for his help, and they set to work together to deal with the emergency. The railway bridge was so badly damaged that it had to be entirely removed, and replaced by a temporary wooden structure. Dargan and his men were encouraged to maximum effort, clearing the debris, driving new piles into the river and erecting the new bridge. The easterly gales went on, and it was necessary to strengthen the parapets on the seaward side of the embankment. In spite of all these difficulties, the repair work proceeded so well that

The *Hibernia* locomotive, made by Sharp, Roberts & Co. for the Dublin & Kingstown Railway. Drawing from *Thirteen Views . . .* , 1834.

LOCOMOTIVE ENGINE—DUBLIN AND KINGSTOWN RAILWAY.

the Directors were able on 5 December to fix the 17th as the date for opening the railway.[6] Advertisements were placarded all over the city, while Vignoles made final trips up and down the line, briefing gate-keepers, police and station staff, until confident that 'after a week's practice everything would be brought into a proper degree of order'.

The morning of 17 December was bright and sunny. At nine o'clock the first trains of passengers started from either end of the railway, and trains continued to run at every hour throughout the day. Vignoles records that upwards of 5000 passengers were carried during the day in 19 trains, of eight or nine coaches, all full to overflowing. At times the crowds burst the barriers and at Kingstown 'an attempt was made on the money-boxes'. According to the Dublin correspondent of *The Times*, Bergin's buffing apparatus was far from perfect, as several gentlemen's heads suffered severe contusions from knocking them against the carriage doors when the trains stopped. (As the majority were Irish heads, *The Times* concluded that they would not mind a few knocks.)

The pamphlet by P. Dixon Hardy, *Thirteen Views on the Dublin & Kingstown Railway*, published on the day of the opening, gives a more poetic account of a first journey on the line.

Hurried forward by the agency of steam, the astonished passenger glides like Asmodeus, over the summits of the houses and streets of our City – presently is transported through green fields and tufts of trees – then skims across the surface of the sea, coasts among the marine villas, and through rocky excavations, until he finds himself in the centre of a vast port, which unites in pleasing confusion the bustle of a commercial town with the amusements of a fashionable watering-place.

Of course the day ended with a sumptuous dinner, given by the Directors for the entertainment of the subscribers, engineers and contractors, at the Salthill Hotel. It was a proud occasion for Charles Vignoles. Happily he could not know until the following morning that Mary had died that day, barely four hours after the first trains had left Dublin and Kingstown.

The three eldest boys, Charles, Hutton and Henry, and a few friends, accompanied Vignoles to the funeral at Liverpool New Cemetery. In a letter to his aunt, Isabella Hutton, he wrote: 'I cannot pretend to great regret, but I cannot forget the former Scenes and Events we witnessed together.' It was some comfort to him that a post-mortem examination had revealed no organic or constitutional disorder of the brain. Mary's mental sickness had been brought about by her own unfortunate temperament, accentuated by the tensions of ill-health and their unhappy married life. Her death could only be a release to them both.

Before continuing the story of Vignoles's work as a whole, it will be convenient to trace his involvement in the developments in Ireland which followed the opening of the Dublin & Kingstown Railway. He was perhaps exaggerating when he told Miss Hutton that the new railway was 'a work perfectly unequalled in Europe', but it had certainly caught the imagination of the public. The Dublin pamphleteer saw in it a means of introducing a general railway system 'whereby the English landlord and traveller may visit the remote parts of Ireland with the rapidity and safety with which he now posts from London to Brighton'. The question was: how was such a system to be established? Vignoles believed that a network of efficient railways would go a long way towards solving the unhappy 'Irish question', and we have seen that he had the vision to appreciate the possibilities of major trunk lines. Early in 1835 he projected a line from Kingstown to Valentia, and incorporated the plan in a map of the whole country, showing a network of main lines, with a table of distances, centres of population and the timing of through journeys from London.[7] On the other hand he was not averse to accepting offers from the various private companies whose formation was being encouraged by the success of the D.&K.R.

Colonel Burgoyne, however, from his experience as Chairman of the Irish Board of Public Works, feared that piecemeal development by private enterprise, on the English pattern, would favour the more prosperous parts of Ireland at the expense of the poorer areas. He was in favour of a system of state control. Largely through his influence a Royal Commission was set up in October 1836, under the chairmanship of Captain Thomas Drummond, R.E., Under-Secretary of State for Ireland, charged with the consideration of a national system of Irish railways. He himself was appointed a member of the Commission, the other two members being Peter Barlow and the Irish engineer Richard Griffith.

By this time Vignoles had become consulting engineer to the Dublin & Drogheda, the Cork & Passage, and the Cork & Limerick companies, and had battled on their behalf in the heavy Parliamentary session of 1836. At the end of May he had also been appointed Chief Engineer of the Great Central Irish Railway. However, this did not stand in the way of his appointment as one of the engineers to the Commission, the other being John MacNeill, an Irishman who had worked with Telford on road engineering. Vignoles was responsible for the south and south-west of the country, MacNeill for the north and north-west. Like Vignoles, MacNeill had surveyed lines for private companies, in the north. No one seems to have questioned the propriety of the Commission's engineers being engaged in private work, although, in a personal letter to Vignoles, Burgoyne expressed some misgivings about his 'coquetting' with other parties.

In addition to its main charge, the Commission was to consider the establishment of lines to ports on the south-west coast suitable for trade with the U.S.A. This must have particularly appealed to Vignoles. He had failed in his attempt to get a trunk route from London to Paris. But what about a through route from London to New York?

He began his work for the Commission in November 1836, having just completed the Parliamentary plans of the Great Central Irish Railway. On 6 December he set out in his carriage on a three weeks' tour to the south, exploring lines previously planned by Griffith from Dublin to Cork and beyond. In a series of long and hastily written letters to the Secretary of the Commission, he recommended various modifications to Griffith's route, particularly to the line onward from Cork to Bantry Bay,

where a packet station for the American traffic was projected at Bereha-
ven. With his usual eye for artistic effect, he envisaged a long viaduct
across the chain of small islands in Glengarriff Bay, which 'would be a
very striking Object in connection with the magnificent Mountain Scenery of the Amphitheatre . . . and would form an additional Object of At-
traction for Visitors'. Throughout the journey his letters speak of appalling weather and of his being perpetually wet through, and he expressed
fears that the peasantry would be in great distress before the end of the
winter.

On 23 December he reported verbally to the Commission on this preliminary survey. The same night he sailed for England to spend Christmas
with his family, after assembling his 'best and most trustworthy assistants'
at Westland Row to receive their orders for taking the levels and sections
of the various lines in detail, a task which was to occupy them for several
months.

The Commissioners issued an interim report on 11 March 1837, and
their final report on 13 July 1838, after both engineers had been required
to lower the cost of their estimates by allowing for a steeper maximum
gradient (1 in 200 instead of 1 in 300). They made out a very reasonable
case for a large measure of public control in the future development of
railways, though there was no question of taking over existing lines. They
pointed to the recent action of the French government in projecting a system of main lines, in contrast with England where main communications
were committed to the almost unconditional and uncontrolled direction
of individuals; and they also pointed to the benefits which could arise
from the employment provided by the construction of railways and other
public works, and from the development of the country's resources by
improved communications.

The Commissioners emphasised the role of a trunk line to the south-
west in providing a rapid rail-service from London to the U.S.A. To this
Vignoles made his own contribution in two proposals embodied in appendices to the report. The first was for a new line (surveyed by him and Rastrick in 1836) from London to a port to be built on the north-west shore
of the Lleyn Peninsula in Wales. With the completion of the London &
Birmingham and the Grand Junction railways the nearest railway port to
Dublin was Liverpool; but the route proposed by Vignoles, though it
would involve heavy gradients in the hills and valleys south of the Snow-
don massif, was more direct, and provided a much shorter sea crossing.
Overall he claimed that it would cut the time taken by a journey from
London to Dublin via Liverpool by four hours.

His second proposal was to complete the coast to coast route in Ireland
by an overhead railway through Dublin. With an eye to preserving the
architectural quality of the city he proposed to carry the railway, not on
brick arches, but on a light iron colonnade viaduct 'of Grecian architec-
ture of the Ionic order' of which the entablature would form a parapet
for the line. Starting from the projected terminus at Barrack Bridge in
the west of the city, the railway would for most of its route follow the
south quay of the Liffey. One row of columns would be on the edge of

Vignoles's proposed
elevated railway
through the city of
Dublin.
Lithograph by G.
Hawkins, from a
drawing by J. D.
Jones.
The view is taken
from the south end
of Carlisle (now
O'Connell) Bridge,
and shows D'Ollier
Street on the left
and Westmorland
Street on the right
with the portico of
the Bank of Ireland
in the distance. The
central block of
buildings is little
changed today. Note
the horse-drawn
train crossing
D'Ollier Street.

the quay, the other would spring from a 'surbase' or wall erected in the river. This, with the aid of a sluice downstream, would maintain the water in a channel beneath the railway at the level of high neap tides. By this means Vignoles hoped to contribute to the salubrity as well as the beauty of the city, for the depth of water contained in the lateral channel would permanently cover the mouths of the numerous sewers which discharged into the river, normally exposed at low water; while at spring tides the lateral channel would overflow and be flushed out by the rapidity of the ebb. Vignoles claimed (a trifle naïvely) that the view from the quayside houses would hardly be affected, since from the lower windows the eye 'could range through the spacious inter-columniations to take in all now seen', while from the upper floors one would be able to overlook the railway parapet. Moreover, lest the citizens of Dublin should complain of noise, the traffic of the railway was to be carried on by horse power, or by endless ropes worked by stationary engines at either end of the viaduct, designed to consume their own smoke. On leaving the quay near Carlisle (now O'Connell) Bridge the line would cross a number of streets, and even pass through rows of houses, but Vignoles calculated that only 37 houses would have to be demolished or materially injured, none of them 'of great value'. In fact the only interference with vested rights would be where the line was to skirt the north wall of the College Park, for 200 yards. But Vignoles hoped to win over the college authorities by providing them with a granite-paved walk beneath the railway which 'would afford, in winter a warm, dry promenade; and in summer a cool

VIEW OF THE PROPOSED
RAILWAY COLONNADE, THROUGH THE CITY OF DUBLIN.

Vignoles's proposed elevated railway along the quays of Dublin.
Lithográph by G. Hawkins from a drawing by Thomas Turner. The view is from Whitworth Bridge, looking towards Usher's Quay, Queen's Bridge and the Wellington Column in Phoenix Park. The last coach of a train on the colonnade is disappearing at the left of the picture.

retreat for the Professors and Students'.

Though the Commissioners were pleased to allow the scheme to be included as an appendix to their report, they did not give it their official blessing. (Vignoles was required to pay the costs of printing his plans.) Only the report and two coloured lithographs survive to show us what might have been. Had Vignoles come to build the line, he would certainly have had to sacrifice some artistic elegance to the interests of structural stability. The picture we are left with, of academic groves under the arches, and contented citizens gazing tranquilly at trains gliding silently past their upper windows, is more a testimony to his artistic imagination than to his engineering skill. As for the transatlantic mail-route, this aspect of the Commission's report had already been forestalled by three months, by the first Atlantic crossing of Brunel's *Great Western*, the forerunner of the liners which were ultimately to render a route across Ireland unnecessary.

The publication of the Commission's report brought a storm of protest. The Press was generally hostile, as were the promoters of private companies which had been unable to complete their plans while the report was being prepared, and now saw themselves threatened by government interference. The supporters of the Great Central Irish Railway were particularly bitter. They accused the Commission of having paid too much attention to the evidence of Mahony, Pim and Vignoles and other supporters of the D.&K.R. Vignoles was said to have deliberately misled

VIEW OF THE PROPOSED
RAILWAY COLONNADE, ALONG THE QUAYS OF DUBLIN.
as it would appear from Whitworth Bridge, looking towards Usher's Quay.
DESIGNED FOR THE PURPOSE OF CONNECTING KINGSTOWN HARBOUR WITH THE GENERAL RAILWAY TERMINUS AT BARRACK BRIDGE.
OF THE VARIOUS LINES FROM THE SOUTH AND SOUTH-WESTERN DISTRICTS OF IRELAND, AS LAID OUT UNDER THE DIRECTION OF THE COMMISSIONERS.
CHARLES VIGNOLES, Civil Engineer F.R.A.S. M.INST.C.E. M.R.I.A.

the directors of the company by encouraging them to believe the Commission would approve their line, and in accepting the post of engineer to the Commission he was accused of playing a double game. The result was that Vignoles resigned from his post with the Great Central, the company refused to pay him for work done after he joined the Commission, and he was obliged to bring a legal action against it to recover what was due to him.

However, not everyone was hostile to the Commission's proposals. There is plenty of evidence to suggest, for example, that the general public in Ireland would have welcomed their realisation in some form or other, as the long period of agitation which followed the publication of the report demonstrated. Such support as there was encouraged Lord Morpeth, the Secretary of State for Ireland, to introduce a Bill in Parliament in March 1839, authorising the government to raise £2½ million for the construction of railways in Ireland. In this he had the backing of those M.P.s who argued that such a development, by helping to reduce the severe unemployment in Ireland at that time, would greatly contribute to the well-being and stability of that country; a point which Vignoles was never tired of making.

The Bill was fiercely opposed by Sir Robert Peel, in the interests of Irish self-reliance and British capital, but managed none the less to get through the first reading in the Commons. Vignoles, who had contributed an article on the subject to the *Dublin Review*, was assured by Burgoyne that he would be appointed Engineer-in-Chief if the project came under the control of the Irish Board of Public Works. For the next two months he took an active part in lobbying M.P.s, preparing papers and plans for public meetings, and briefing the Secretary of State for the Bill's next stage. But on 21 June, to his great mortification, the Cabinet decided not to proceed with the Bill, owing to Tory opposition and the lack of support from ministers. 'So ends the labor and expense of several years!!!' was Vignoles's indignant comment in his diary.

There was to be one more attempt a couple of years later to introduce a state-controlled railway system in Ireland, but it was equally unsuccessful. Unhappily for that country, valuable time had been lost during the period of the Commission's work, during which private enterprise was discouraged from launching new schemes. Meanwhile money was being widely invested in England. When the way was finally clear for the private promoter in Ireland, investors hung back, reluctant to venture into what appeared to be an unprofitable field of investment in comparison with England. This was, of course, exactly what Burgoyne, Drummond and Vignoles had foreseen, and what they had hoped to forestall by the development of a state-controlled system.

Thus the country fell between two stools. Such railways as were built by private enterprise proved to be inefficient and unremunerative, and within 20 years or so there were widespread demands for amalgamation and state support.

6 The North Union, the Midland Counties railways, and the Vignoles rail, 1834–1839

During the events in Ireland described in the last chapter, Vignoles's other commitments continued to increase. In November 1833 he completed the Parliamentary plans for the London to Brighton line. In 1834 he personally carried out preliminary surveys for three projects in different parts of the country. The first, on behalf of the Thames & Severn Railway Company (an extension of the London & Windsor), took him by gig, on horseback and on foot along the Thames valley, and over the Berkshire Downs, exploring an alternative line to I. K. Brunel's proposed Great Western. The route he chose was to run north of Brunel's line through the Vale of White Horse, and climb through the southern Cotswolds to reach the Severn at Sharpness. Here he envisaged an ambitious 20-arch viaduct to carry the line into South Wales, and branches to Gloucester and Bristol. Following his visit to Kensington Palace in July, he travelled to North Devon, where two of his assistants, Terry and Talbot, were already at work on a survey for a ship canal from Barnstaple, to by-pass the winding channel of the River Taw estuary. Vignoles spent two long days walking over marshes and sand dunes, and in pilot boats up and down the river, closely studying tides and channels, before presenting outline plans to the promoters of the scheme; after which, having instructed his assistants to complete the survey by the end of the month, he returned by coach to Bristol, where he embarked for the 36-hour passage back to Kingstown.[1]

In October of the same year, he assisted John Braithwaite in a rapid exploration of the route proposed for the Eastern Counties Railway from London to Colchester and Norwich. The promoters of the E.C.R., whose secretary was J. C. Robertson, editor of the *Mechanic's Magazine*, were anxious to have the plans ready for the coming Parliamentary session, but this proved to be beyond even Vignoles's resources, with the final work on the Dublin & Kingstown Railway on his hands. In the end the bulk of the work on the Eastern Counties was to devolve on Braithwaite, with Vignoles in the background as consultant. All these schemes naturally entailed hours of office work after completion of the surveys, preparing plans and reports.

As was to be expected, the Thames & Severn Company were leading opponents in the Parliamentary hearings of the Great Western Railway Bill. Vignoles was called as an opposition witness, and, according to his own account, advised counsel on technical points in their cross-examination of Brunel, and even went so far as to prompt them in committee. It appears however that the latter were so slow in grasping engineering technicalities that he lost patience with them, and, as he wrote in his diary, 'in consequence of the perverse manner in which Counsel acted I refused to interfere further, and resolved to leave London until it came to my turn to be examined'. Vignoles's enthusiasm must sometimes have been the cause of some embarrassment to his friends.

The G.W.R. Bill was not finally passed until the following year, by which time the Thames & Severn line had faded out, and Vignoles had seized the opportunity of a reconciliation with Brunel, by giving evidence in his support. In 1834 he was also supporting Locke and Stephenson in the committee stage of the Grand Junction Bill (noting in his diary on 15 April 'Consultation with Mr George Stephenson!!!'). In the same session the new version of the Dublin & Kingstown Extension Bill, and the North Union Bill, had received the Royal Assent, the latter event resulting in the confirmation of Vignoles's appointment as Chief Engineer of the North Union Railway at a salary of £1000 per annum.

The prospectus of the North Union Railway was issued in October 1834. The southern section of the line – the Wigan Branch Railway – was already built. The contracts for the section from Wigan to Preston were let in the following January. It was a straightforward line, with a rather stiff gradient over Coppull Moor, at the end of which it had to be carried over the modest ravine of the River Yarrow, before running down to cross the Ribble immediately south of Preston, where Vignoles planned a stone bridge of five elliptical arches. (See map, p. 27.)

The early work on the North Union Railway took place at the busiest period of the D.&K.R. construction, and it was difficult for Vignoles to give it all his attention. A great deal of responsibility was to be carried by two of his most experienced assistants, William Coulthard and John Collister. He was never afraid of delegating responsibility to those he employed. He gave them meticulous instructions, made periodic inspections, and otherwise left them to work with as little interference as possible; and they gave him efficient and loyal service, often for many years. Yet, in spite of his extraordinary capacity for dealing with several different undertakings simultaneously, it was perhaps inevitable that he should on occasions overreach himself, simply through the impossibility of being in two places at once. One such occasion, when he failed to keep an appointment to meet the North Union Railway Directors on the ground, brought a kindly but firm rebuke from Theodore Rathbone, Vice-Chairman of the Company. Writing as a 'very sincere well-wisher' he advised Vignoles to beware of trying to undertake too much. He went on:

Your talents and energies are known to be of no common order, but still you have bitter ennemies [sic] . . . and you cannot be aware how hard the occurrence of instances of rashness and thoughtlessness of this kind make it for your friends to fight your battles . . . if you have too much on hand to comply with this engagement on its present footing, it would be better far to reconsider it and frankly communicate the result to the directors.

It is hard to realise that the man to whom the above was written was 41 years old! Vignoles retained well into middle age the rashness and impetuosity of youth, as he also retained its vigour, enthusiasm and self-confidence. There is nothing diffident or timid in the face that looks out at us from the Roffe engraving of 1835, and it is that of an essentially young man.

Whether he paid heed to Rathbone's warning or not, he showed no sign of trying to limit the scope of his work. Hardly was the Dublin & Kingstown line open when he was in North Germany, examining the possibility of a line from Hamburg to Hanover and Brunswick, which had been mooted in London business circles early in 1834.

Apart from the time spent on the survey, Vignoles was at pains to enlist local support for the project. He countered criticisms of his plans by the publication of a pamphlet illustrated with an engraving of a locomotive and carriages from the *Railway Companion*. He spoke in French at a public meeting of merchants at Hanover, giving them 'a popular description of the Railway, etc.', and at a dinner given by the railway promoters in Hamburg, when he expounded his ideas for the bridging of the River Elbe and the improvement of Hamburg Docks. Most important of all, while in Hanover he had an audience of the Duke of Cambridge, at that time regent for the young Prince Ernest of Hanover, and secured his support for the railway. It was on this visit that Vignoles first met Herr Hubbé the Hamburg city engineer, with whom he formed what was to be a lasting and fruitful friendship.

He returned to England on 21 March, and reported to the London committee four days later. (Characteristically he records in his diary a meeting with Braithwaite to discuss the best method of heating the royal palace at Brunswick!)

The railway was ultimately built by German engineers, though Vignoles was consulted frequently over the next few years concerning the plans, as well as the supply of engines, rolling-stock and rails, which were all ordered from England, mainly from Forrester's Vauxhall works in Liverpool. He also collaborated with Herr Hubbé in plans for extensions to the Hamburg Docks.

In the summer of 1835, encouragement to embark on a new project came from Rathbone himself. He was one of a group of Liverpool men on the Board of the proposed Midland Counties Railway. Another was James Cropper, a keen supporter of Braithwaite and Ericsson. The first surveys of the Midland Counties line had been made by George Rennie and William Jessop, but Rennie seems to have lost interest and Jessop had retired to a place on the Board. It was not surprising therefore that the Liverpool men should turn to Vignoles, who in September 1835 was invited to report on the plans.

Proposals for the new line had arisen in the first place as the result of competition between the coal owners of Derbyshire and Nottinghamshire and those of Leicestershire. Until 1832 Leicester's coal supplies had come by canal from north of the Trent; but by the opening that year of the Leicester & Swannington Railway, a cheaper and easier source of supply was provided from the local collieries of Charnwood Forest. The Notts. and Derby coal owners therefore proposed to build a railway from Pinxton down the Erewash Valley and across the Trent to Leicester. In doing this they naturally met with the opposition of both the Leicester & Swannington Railway Company and the canal companies.

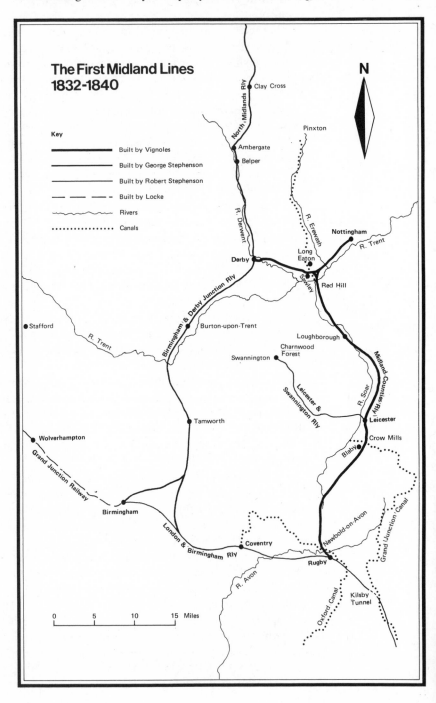

The First Midland Lines 1832-1840

N

Key

——————— Built by Vignoles

——————— Built by George Stephenson

——————— Built by Robert Stephenson

– – – – – Built by Locke

～～～～～ Rivers

················· Canals

Clay Cross

Pinxton

North Midlands Rly

Ambergate

Belper

R. Derwent

R. Erewash

Nottingham

R. Trent

Long Eaton

Derby

Sawley

Red Hill

Birmingham & Derby Junction Rly

Stafford

R. Trent

Burton-upon-Trent

Loughborough

Charnwood Forest

Swannington

Leicester & Swannington Rly

R. Soar

Midland Counties Rly

Leicester

Tamworth

Crow Mills

Wolverhampton

Blaby

Grand Junction Railway

Birmingham

London & Birmingham Rly

Coventry

Newbold-on-Avon

Grand Junction Canal

Rugby

R. Avon

Oxford Canal

Kilsby Tunnel

0   5   10   15 Miles

When the Midland Counties Board was strengthened by a group of Lancashire directors, their chief interest was to improve communications with the south, with a view to capturing the through traffic from Derby and Nottingham to London. The line was now planned to link these two towns north of the Trent, and to run south through Leicester, to join the London & Birmingham line at Rugby. The promoters must also have had their eyes on a rival project further west, where George Stephenson was planning the Birmingham & Derby Junction Railway; for this too was intended to secure the through traffic from London. Thus, when Vignoles became engineer to the Midland Counties line at the end of 1835, he found himself once more in competition with Stephenson.

It was typical of the cut-throat competition of the railway age that these two lines were to obtain their Parliamentary acts within a month of one another, and only a few months would separate their dates of opening. They were then to compete fiercely for the through traffic for over four years, until they were amalgamated as part of the Midland Railway under the dictatorial rule of George Hudson, the 'Railway King'. Included in this merger would be the North Midland, another Stephenson line carrying the through traffic northward from Derby.

Meanwhile, in the last months of 1835, Vignoles was invited to make preliminary surveys on yet another line, the proposed Sheffield, Ashton-under-Lyne & Manchester Railway. His association with this project was to have an important bearing on his whole career, since after offering great opportunities of success, it brought him to the edge of financial disaster. It will be convenient to complete the account of the North Union Railway and the Midland Counties Railway, and reserve that of the Sheffield & Manchester Railway for the next chapter, while remembering that for Vignoles all three projects, as well as his other work, competed simultaneously for his attention.

On the North Union Railway bad weather and difficulties with contractors were a continual source of delay. At the beginning of 1837, the contractor responsible for the southern end of the line, William Hughes, was unable to pay the wages of his workers and the contract had to be taken out of his hands. Later in the year Mackenzie, the holder of the Yarrow contract, was also in trouble. At the end of October the centre of one of the culverts over the river Yarrow collapsed suddenly, blocking the course of the river. Heavy rain caused the water to rise rapidly, and although Vignoles immediately ordered a channel to be dug to carry off the flood, this channel was also blocked by a fall of earth; the water continued to rise in the narrow steep-sided valley until it reached the top of the embankment 30 ft above the river-bed, when it swept away a large section of the embankment and culvert, the valley beyond being buried deep in earth and bricks. Vignoles records that the Directors decided to hold Mackenzie responsible for the damage, 'which resolution being conveyed caused him to give way to the most violent passion and language'. Only after Vignoles had argued the contractor's case with the company solicitor did the Board agree to exonerate the contractor from blame for the damage caused by the flood. Mackenzie also agreed to pay half the cost of the timber viaduct Vignoles designed to replace the wrecked embankment and culvert.

At the northern end of the line there had also been delays. The foundation stone of the Ribble Bridge had been laid as long ago as September 1835, but two years later construction was still moving slowly owing to a shortage of timber for the centrings of the arches. However, on 14 April 1838 the *Railway Times* was able to report that 120 to 130 men were working on the bridge, three of the arches being nearly complete. About this time Brunel was in trouble with one of the two spans of his brick arch across the Thames at Maidenhead; and it is interesting to note that Vignoles visited it on 12 June and recorded his opinion that it would have to be rebuilt. In this he was mistaken, although the failure of a bridge was not an unusual event in the early days of railway engineering. Whether he had any qualms about his own work we do not know; but on 10 September he expressed his satisfaction that the last arch of the Ribble Bridge was 'entirely keyed with the exception of 3 or 4 stones'. It was his most ambitious construction to date, and one that survives to this day, although modified by widening of the line. In its original form it had a total length of 872 ft, five elliptical arches of 120-ft span with a rise of 33 ft, was 68 ft high and 28 ft wide. 675 000 cu. ft of rusticated ashlar, supplied by Whittle & Longridge of Lancaster, was used in its construction, at a total cost of £40 000.

Heavy rain early in October, which threatened the stability of the new embankments, caused further delays. But on 29 October Vignoles brought the old *Novelty* out of retirement and rode on her from Wigan to Preston. He was accompanied by his son Hutton, then 14 years old, who was allowed to drive the engine over the Ribble Bridge, to the shouts

Bridge over the River Ribble at Preston, North Union Railway, 1862. A photograph by the Preston Camera Club.

and huzzas of the numerous workmen on the spot.[2] Vignoles returned with a train to Parkside, where he picked up Theodore Rathbone for a final inspection of the line. On 31 October the first public train left Preston at 6.45 a.m., and a regular service to Liverpool and Manchester via Parkside was maintained all day. Vignoles noted that the trains were 'keeping their Time along the North Union Line very well – but detained at Parkside for want of better arrangements for Coke and Water, a fault which will soon be properly remedied'.

The *Preston Pilot* commended the Mayor of Preston 'who kept a hospitable open house on the occasion' and 'had a beautiful flag floating in the breeze on a staff fixed on his own residence'. It also published interesting details about the time-table and fares. First-class trains made up of carriages carrying six passengers *inside* would only stop twice between Preston and Parkside, while mixed trains of both first- and second-class carriages would stop at all stations. The through single fare from Preston to Manchester or Liverpool was 7s 6d first class, 5s 0d second class, children under ten 4s 0d and 2s 6d. The fare for a horse was 12s 6d, with reductions for two or three horses travelling in the same box and belonging to the same owner.

The new line had been building for nearly four years. Much work was still to be completed at Wigan and Preston stations,[3] and the contractors' final accounts were a matter of considerable dispute with the company, in which Vignoles was required to act as arbiter. But another link had been forged between London and the North. South of the Liverpool & Manchester Railway, plans were well advanced for connecting with the Grand Junction and the London & Birmingham lines. So much so that within a year the Company's minute book would be recording complaints

Original stonework on the west side of Vignoles's Ribble Bridge, photographed in 1973. The bridge has been widened twice since it was built, first by doubling the brick and masonry construction, on the east side, and then by adding a trussed girder bridge on the west side. The shadow of this can be seen in the upper part of the photograph, which is taken from below the extension.

of passengers having to 'shift carriages' so often between Preston and London, and being unable to book right through. Vignoles was retained as consulting engineer at £50 per annum, and Collister and Coulthard as residents until 1841.

Meanwhile, further south, the Midland Counties line had been making steady progress, under the direction of Thomas Woodhouse, the resident engineer. At the northern end the most important feature of the line was again to be a bridge, this time spanning the Trent and its broad water-meadows south of Long Eaton and close to the village of Sawley. Work on the bridge began in June 1838. In contrast to the Ribble Bridge, it was to be a cast-iron structure; when completed it would consist of three graceful arches, each of 100-ft span, mounted on stone piers 40 ft long, 10 ft wide and 22 ft above the level of the water. From the north the bridge was to be approached by ten 25-ft flood-arches built in brick, and on the south two more flood-arches would carry the line into a tunnel 133 yards long through Red Hill, a steep and narrow sandstone ridge flanking the river's southern bank.

William Mackenzie, who had formerly been an assistant to Telford, was the contractor responsible for both bridge and tunnel, the entrance to which is a castellated stone-faced horseshoe in the Gothic style, which can still be seen lurking in the trees on the north side of Red Hill. The handsome and delicately ornamented brickwork of the flood-arches still spans the meadows, but the main structure of Vignoles's bridge, cast to his design by the Butterley Iron Company, has been replaced by a hideous girder construction.[4]

South of the Trent as far as Leicester no gradient on the line exceeded

Cast-iron bridge over the River Trent, near Sawley, Midland Counties Railway, 1840. Lithograph by G. Hawkins Jnr. The view is from the south bank of the Trent looking east, and shows Red Hill Tunnel, and the junction of the Soar and Trent.

1 in 500, but in the upland country between Leicester and Rugby heavy earthworks were necessary to maintain a steady climb over a summit of 407 feet, from which the line ran down for five miles at a gradient of 1 in 330 (the maximum fixed for the London & Birmingham Railway), the last half mile into Rugby being on the level. Two viaducts were to be built on this section, one near Newbold-on-Avon, over the Avon and the Oxford Canal, the other at Crow Mills, crossing the Grand Union Canal at Blaby.

At the time of the building of the Midland Counties Railway, Vignoles was much concerned with the question of improvements in the materials and design of permanent way. Following the example of George Stephenson he had laid the Dublin & Kingstown Railway tracks in chairs mounted on stone blocks. On the North Union, malleable or wrought-iron rails rested in chairs at the joints, the rails being keyed into the chairs with 2½-in. oak wedges, the chairs being fastened to the stone blocks by spikes driven into oak plugs, inserted into holes 1¼-in. in diameter and 6 in. deep. But he was now coming to the conclusion that the excessive rigidity of this form of support was causing severe wear and tear not only to the rails, but also to the wheels and frames of engines and rolling-stock.

Red Hill Tunnel, Midland Counties Railway, *c.* 1950

He was to sum this up in his report to the Irish Railway Commissioners in 1838 when, referring to the Dublin & Kingstown line, he wrote of the 'continued contest between the infringing action of the Trains and the *vis inertia* of the Railway, the resistance of the blocks of this non-elastic road creating a much larger amount of positive wear than was suspected both to the Engines and the Carriages; and hence arose the mutual destruction of the Railway and the Vehicles moving on it'.

It is possible that it was a meeting between Vignoles and the U.S. engineer R. L. Stevens, some time in the winter of 1830–31, that set him thinking about using a flat-bottomed rail which would bear directly on the sleeper, without the need of a chair. Stevens, an engineering colleague of Major Wilson, Vignoles's former chief in Charleston, was building a line between Amboy and Camden in the United States, and according to Auguste Moyaux, a Belgian historian writing in 1905, he devised a new type of flat-bottomed rail, and had it rolled in England. Moyaux goes on to say that Vignoles introduced this rail on the continent and gave it his name.[5] Whether Vignoles had the idea from Stevens, or Stevens from Vignoles, does not perhaps matter. What is important is that this was the first appearance of the flat-bottomed rail of inverted T section, fastened directly to the sleepers with clamps or broad-headed spikes, which is known in France as 'le rail Vignole' (*sic*) and in Germany as 'die Vignolesche Schiene'.

As Vignoles was one of the first engineers to be consulted by French and German railway promoters there seems no doubt that he recommended this type of rail to them; indeed, writing on 1 December 1843, Vignoles records that a German engineer in Württemberg showed him a drawing of a proposed rail which was an adaptation of a form of rail 'originally proposed by me in the year 1830 and which are well known in all Germany as the "Vignolesche Rail"'. (In giving the date as 1830, Vignoles was perhaps thinking of his first recommendation of the rail in England, rather than in Germany.)

His countrymen were more difficult to persuade. In 1835 Robert Stephenson was still using, for the London & Birmingham Railway, the fish-bellied rails and stone blocks his father had laid down on the Liverpool & Manchester Railway. Vignoles, on the North Union Railway, had moved to a rolled-iron rail of double T section weighing 55 lb per yard, equal top and bottom, but still fastened with chairs. In May 1836, however, in his capacity as consulting engineer of the London & Croydon

Typical sections of different rails.

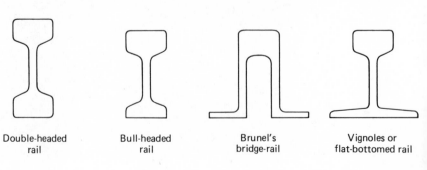

| Double-headed rail | Bull-headed rail | Brunel's bridge-rail | Vignoles or flat-bottomed rail |

Railway, he persuaded the Directors to adopt a flat-bottomed rail, laid without chairs on longitudinal baulks of timber 8 in. by 5 in. in section and tied transversely with wooden sleepers. The continuous support given by the longitudinal timbers was an innovation by means of which Vignoles hoped to achieve a smoother ride, while the substitution of timbers for stone would, he claimed, provide greater elasticity.

It is interesting to note that about this same time I. K. Brunel was planning to use longitudinal timbers on his Great Western Railway, to support the 'bridge-rail' of inverted U section which he had evolved. His timber supports were also united with cross ties, making a solid timber frame as a bed for the road; and thereby helping to preserve the gauge, which in the case of Stephenson's track needed continual maintenance.

Vignoles repeatedly urged his masters on the Midland Counties Board to give his rail a trial, with the result that in January 1837 they agreed to allot £500 to an experiment to be conducted, not on their own system, but on the section of the London & Birmingham line which was due to be opened in June of that year, provided that company would allow the experiment.

It seems to have been left to Vignoles to make the approach, as it was he who reported to the Board on 16 June that the London & Birmingham Railway had agreed to lay 534 yards of experimental line near Harrow. No doubt the friendly relations he had now established with Robert Stephenson helped to smooth the way. At about the same time he was recommending the re-laying of a mile of the Dublin & Kingstown Railway on the same principle.

The result of these experiments convinced him that the system had the advantages he had claimed for it, being much less liable to wear and tear, besides being cheaper to lay and maintain. However the Directors of the Midland Counties Railway were not to be persuaded. They did make the concession of laying a meagre five miles of their line with Vignoles's rails. Otherwise they persisted in adopting the double-headed rail of the type introduced by Joseph Locke on the Grand Junction line. The rails were 15 ft long, weighing 77 lb per yard, and were laid in chairs, on blocks of Derbyshire millstone grit, with the result that in a very few years the line had to be relaid on timber, many of the blocks ending their days built into the platforms of railway stations.

One of the arguments put forward in favour of the double-headed rail was that it could be turned over after several years of use, thus prolonging its working life. In fact this did not prove practicable, as it was found that the lower head of the rail was worn where it had rested in the chair. But it was the forerunner of the so-called 'bull-headed' rail, in which the head and foot were of the same width but the head was deeper. The foot became a convenient flange for keying the rail into its chairs. The bull-headed rail was to be generally adopted on British railways from about 1860 onwards, and in 1905 it became the British standard, at a weight of 95 lb or 85 lb per yard. The 'fishing' of rails (the end-to-end joining of rails by means of 'fishplates' bolted together on either side of the rail) had been general practice from an early date.

The engineers continued to be divided in their opinions about the merits of the various types. At a discussion of a paper read to the I.C.E. by Francis Fox in February 1861, Vignoles stressed the importance of elasticity, and declared a preference for Vignoles rails laid on cross sleepers, while G. P. Bidder, the President, preferred the double-headed rail on cross sleepers 'now almost universally adopted'. A substantial paper read to the I.C.E. by Price Williams on 20 March 1866, 'On the Maintenance and Renewal of Permanent Way', provoked a discussion which was extended over the next three meetings of the Institution. In this discussion Vignoles pointed out that the double-headed rail was liable to snap when reversed,[6] and claimed that there could be no improvement in the permanent way so long as chairs were used. It was remarkable, he said, that every country used the flat-bottomed rail except England. Rails weighing 80 lb per yard and chairs of 35 lb each could be replaced by a flat-bottomed rail of 100 lb per yard without chairs. The 'fishing' of rails produced the effect of a continuous rail, and helped to cut down the dislocation which occurred at every joint when a wheel passed over it. Vignoles was strongly supported by G. W. Hemans, who pointed out that the Vignoles system transferred the weight of the locomotive directly to the ballast, and was therefore more suited to sustain heavy weight than a system of supports at short intervals. He claimed that the heavy Vignoles rail, well fished and laid on longitudinal sleepers, had never really been tried in Great Britain.

John Fowler, the President, summed up a long and wide-ranging discussion by pointing to the successful use of the Vignoles rail in France, where traffic was as heavy and speeds as high as any in Great Britain.[7] He concluded that the best type of permanent way had not yet been evolved.

As far as British railways were concerned, it was to take nearly 90 years for the wheel to come full circle. Despite its advantages as regards cost of maintenance, ease of laying, strength and distribution of weight, the Vignoles rail was harder to roll, and the spikes with which it was fastened were not always adequate. They were as rigid as the chairs they replaced, and liable to break; and they failed to give the strong lateral support provided by the chair system. While the substitution of steel for iron, during the latter part of the century, and the steady improvement in the technology of rail production, made for greater strength, the increase in speed and axle-load during the same period intensified the three-way movement, lateral, longitudinal and vertical, transmitted to the rails by the passage of the locomotive and rolling-stock. To cope with this a more resilient or elastic type of fastening was badly needed for use with the flat-bottomed rail.

The way to this was finally opened, with the invention in 1934, by the German engineer Max Rüping, of a resilient spring-headed rail-fastening, and its development during the Second World War by the Elastic Rail Spike Company, an Anglo-American venture incorporated in London. With this invention, and its successors, the Lockspike, and the Pandrol Clip, the newly nationalised British Railways were ready in 1949 to settle for the general introduction of the flat-bottomed rail throughout the British system, with rails weighing 109 lb and 98 lb per yard. Some dozen years later, the introduction of continuous rail welding as a substitute for 'fishing' was a further step forward. The Vignoles rail was finally justified.

To return to the year 1839, the first section of the Midland Counties line between Nottingham and Derby was opened on 20 May that year; the great bridge over the Trent was completed that autumn, and the remaining sections of the line were opened on 4 May and 29 June of the following year. George Stephenson's Birmingham & Derby line had been open since August 1838. At Derby he and Vignoles had agreed to the building of a joint station, the cost being borne by the North Midland Railway, the other two lines paying a rent of six per cent on their share of the accommodation. The building of the Midland Counties line had caused no great anxieties to the engineers, but a dramatic epilogue to its construction was provided by the collapse in late November 1840 of one of the piers of the Crow Mills Viaduct, luckily without loss of life or injury. Vignoles attributed the accident to the use of Barrow Lime in grouting the foundations, an unsuitable material which had failed to cement, and he placed the liability firmly on the contractor.

Serious as this disaster might have been, it was small compared with the personal difficulties which were now overwhelming Vignoles through his association with the Sheffield & Manchester line. But before taking up the story, we must add a brief note on family history.

Since the death of his wife he had done his best for his children's welfare. Though he did not spoil them, he did not grudge them what he thought was necessary for their maintenance and education. He had taken young Charles to Germany and left him with a tutor in Brunswick, Hutton was at school in France, Henry and Olinthus alternating between the care of a friend at Peckham and a small preparatory school. Only Camilla was giving him cause for unhappiness. After her return from France she had fallen in love with a young man called Thomas Croudace, who in the spring of 1837 asked permission to marry her. Vignoles was not pleased with the idea of such an early marriage when it was put to him, nor with the proposed husband, who worked at the Coal Exchange. He removed Camilla to a safe distance from London, first to stay with his cousin Henry Hutton, Rector of Woburn, then to board in a private house in Thames Ditton. Faced with her father's disapproval, Camilla showed whose daughter she was. On 31 May, her father's forty-fourth birthday, she was secretly married to Thomas Croudace. The couple eloped from Thames Ditton on 10 June. It was not until 16 June that Vignoles heard the news and made a brief unemotional note of it in his diary.

Not long afterwards a friend in Paris, who knew Camilla well, wrote to Vignoles: 'Poor thing, She little reflected on the misery that almost always follows marrying contrary to the wishes of a parent, and one she so much respected and loved.' One wonders what sort of a chord of memory this comment struck in her father's heart.

7 The Sheffield & Manchester disaster, 1835–1840

The Sheffield & Manchester Railway was a boldly conceived plan to bring the benefits of the railway system of Lancashire to the merchants and industrialists of Sheffield and Yorkshire, by carrying a line over the main ridge of the Pennines a few miles north of the Peak. It presented an engineering problem of an entirely new order, as Vignoles discovered when he made his first exploratory survey of the high Pennine moorland during the short December days of 1835. Seven miles north of the Peak the summit of the ridge is at the lowest point, 1400 feet, approached on the south-west by the deeply cut valley of the River Etherow, and on the north-east by the more gentle slope of the River Don. Vignoles planned to bring his line along these two valleys, driving a tunnel to join them beneath the desolate moorland hump where the Manchester–Sheffield turnpike road heaves its way over the ridge. He presented his plans at public meetings on 4 and 5 January 1836, one at Sheffield, the other at Manchester. At Sheffield, Lord Wharncliffe, a local landowner and industrialist who was ultimately to be chairman of the company, presided and both meetings warmly supported the principle of the railway, voted for the formation of a company, and approved Vignoles's plans. These took into account the possibilities of junctions at either end of the line, with the L.&M.R. at Manchester, and with the North Midland Railway at Sheffield. (See map, p. 27.)

Critics of the scheme were of course not lacking, and among them were those who were daunted by the sheer magnitude of the operation. The tunnel alone would be in a class by itself; driven through the millstone grit of the Pennines, it was to be over 5000 yards long, compared with three other tunnels under construction about this time: Robert Stephenson's Kilsby Tunnel on the London & Birmingham Railway (2420 yards); Brunel's Box Tunnel on the Great Western Railway (3230 yards); and George Stephenson's Littleborough Tunnel on the Manchester & Leeds Railway (2869 yards). Such misgivings must have prompted the decision of the provisional committees (there were two of these, one at either end of the line) to engage Joseph Locke, at this time engineer of the Grand Junction Railway, in addition to Vignoles, and to invite both engineers to make separate surveys and estimates. Vignoles records in his diary a protest made by him in June, which may well have been in connection with this arrangement, but otherwise he seems to have gone steadily ahead with the survey through the summer, posting from end to end of the line, trudging up and down the long valleys with his assistants, clambering over rock and heather, sometimes drenched in torrential rain, drafting, revising, discussing with the Directors his selection of curves and gradients. On the Manchester side the configuration of the ground would necessitate the building of a viaduct across the Etherow at Broadbottom, near Dinting Vale, and another across Dinting Vale itself, in order to maintain the height necessary for a steady rise of 450 feet in the seven miles from the Etherow to the tunnel face. On the Sheffield side the descent was more gentle, as it followed the Don valley.

After a meeting on 17 August to compare notes, the two engineers submitted their separate reports, which were considered by the management committee on 14 October. There was little difference between the two plans, though Vignoles was informed in confidence that his was preferred by the Directors. An agreed plan was eventually produced, after the Chairman had been obliged to intervene to persuade Locke to confirm Vignoles's report. Olinthus Vignoles interpreted this incident as symptomatic of the continued friction arising from the existence of the rival committees, a division which was to bedevil the history of the line, and to cause considerable embarrassment to its Chief Engineer; for although Locke was now content to leave the field vacant to Vignoles, who was generally approved by the combined management committee, a minority on the Manchester committee would have preferred to retain Locke, and took every opportunity of putting difficulties in Vignoles's way.

The Sheffield & Manchester Bill passed the House of Lords in April 1837, but unfortunately for the new company the year brought the beginning of a trade recession. A run of good harvests had come to an end, and there was somewhat of a financial crisis, stemming from strains in the United States. The Company had voted a capital issue of £1 000 000 in £100 shares. According to the existing law, shareholders were required to subscribe five per cent of the value of their shares in the first instance, with a promise to pay the balance in instalments, when called upon. But, although the Parliamentary Bill was obtained in April, subscribers were slow in coming forward. Vignoles, who was itching to get on with the work, endeavoured to inspire the subscribers with confidence by the issue on 20 May 1837 of a detailed report 'on the mode of commencing the works'. While presuming that it would be 'an object with the proprietors [i.e. shareholders] to authorise the Directors to pursue with vigour the execution of the Works', he recommended that it was necessary to allay the mistrust which

will naturally arise in the minds of persons unfamiliar with Public Works on a large Scale, who having no standard of comparison in their own minds whereto to refer the projected measures, will often condemn them as impracticable. In the case of the Sheffield & Manchester Railway some points of this kind have been presented, of which the principal, as regards the execution of the work, are the Tunnel through the Summit ridge, and the Viaducts at Dinting Vale & the Etherow River. The rest of the physical difficulties are not greater than have been overcome in other places with facility and success.

Among the points made by Vignoles to reassure the faint-hearted Directors and shareholders were the following: The Engineer must exercise due economy. The gradients would not exceed 40 ft to the mile (1 in 132) and locomotives would be used throughout. At least 12 months would be devoted to the survey and laying out of the line, during which time the strata of the summit ridge would be thoroughly explored 'by active and decided measures and at moderate expense'.

He proposed to sink 12 shafts eight to nine feet in diameter at quarter-mile intervals along the line of the tunnel. This would provide 24 working faces from which driftways would be driven, each 220 yards long, entered laterally from the foot of each shaft and driven alongside the centre line of the tunnel. He considered there would be no difficulty in finding enough working miners for all the shafts to be worked simultaneously, thereby ensuring completion of this stage in 12 months.

Once this exploration was complete the main tunnel would be driven from the shafts, while the drifts would provide a drainage system. At first only a single line of railway would be provided for, to be completed in the second year. A second would be constructed as necessary. While the miners were at work the rest of the line would be marked out and quarries opened up in the hills above each of the viaducts, with inclined planes to bring down the stone to the site.

Vignoles recommended that contracts for the work should be allocated in small 'portions', and he emphasised the desirability of pushing on with the staking out of the whole line, taking advantage of the summer weather; he maintained that the balance of funds in hand was sufficient for the exploratory work and the staking out of the railway, and concluded his report with an echo of General Wolfe at Quebec: 'I shall consider it a source of higher fame to have executed the Works of the Railway . . . *economically* than to have designed them.'

The style of this report, and the emphasis Vignoles lays in it on economy, show clearly that it was intended mainly as a boost for the morale of the Directors and shareholders. As a detailed plan it leaves many questions unanswered. But in spite of his efforts to inspire enthusiasm among shareholders subscribers hung back, and some were even seeking to dispose of the shares they had already bought. On 23 October a shareholders' meeting (in Sheffield) resolved to postpone all work, and all further calls on the shares, until the following spring.

The position was critical. For by law, until the shares were fully subscribed for, the compulsory powers of the Act could not be invoked, and eventually the Act itself could become null and void. With the fortunes of the Company at such a low ebb, the time had come for desperate measures. Vignoles had already bought some shares for himself, including a block of a hundred acquired at the beginning of the year in the name of Miss Hutton. He now set about buying up a large number of the depreciated shares – they could be bought for an initial payment of 20s 0d at the end of 1837 – and with the agreement of the Directors placed many of them in the names of friends and relatives, on condition that the shares would be considered as held in trust for him, without the holders incurring any liability for payment on future calls. By this action he hoped to forestall any uneasiness which might have been caused in the public mind by the fact of the engineer holding so many shares. At the same time he canvassed widely for new purchasers, and sought an interview with Lord Wharncliffe, to persuade him of the necessity of taking a much more active interest in the Company's affairs.

In thus identifying himself with the fortunes of the Company, Vignoles was taking a bold, if not a rash, step. But it was one typical of the man, and he was encouraged by his confidence in his own powers to complete the job, as well as in the Directors' promises, by the punctilious way in which he had been paid for the work he had already done. With the amount of work he had on hand elsewhere he was in a stronger position financially than he had ever been before. Admittedly his plans for a London to Brighton line had been withdrawn in the face of the proposals of four other engineers, among them Robert Stephenson and John Rennie, whose line was eventually preferred in Parliament; but he was much in demand as a Parliamentary witness, always a substantial source of fees. Thus he could afford to invest heavily in the prospects of the line he hoped to build, confident that its success would enable him to sell many of his shares before they had to be fully paid up.

As he had hoped, in the spring of 1838 the rate of subscription to the Company's shares began to recover; and by the beginning of March it had increased sufficiently for the Directors to decide on staking out the line. Vignoles was formally appointed engineer on 18 April, undertaking to complete all the necessary plans and documents by 1 June 1839. For this he was to receive a gross sum of £5000 payable in quarterly instalments. The Company undertook to pay the salary of a resident engineer, otherwise he was to be responsible for the salaries and expenses of his assistants.

In June 1838 the Jewish financier Moses A. Goldsmid agreed to act as the Company's London agent, and he and Vignoles prepared a prospectus for publication. In September Messrs Smith, Eckersly and Worswick accepted the contract for the western section of the line, which was already being staked out according to the Parliamentary plans by Vignoles's assistants, Terry, Seed and Wellington Purdon, the last-named being responsible for the line of the tunnel. Vignoles refers in his diary to the interest shown by the Directors who lived near the route, such as the brothers Sidebottom, the proprietors of a large mill at Mottram, and James Rhodes of Tintwhistle. For instance the latter spent a whole day walking over the line of the tunnel with him, discussing the positions of the first shafts and boreholes.

It was important for Vignoles to have a base near the main scene of operations in this wild and desolate country, from which he could also reach Preston, Derby and Liverpool without difficulty. For this purpose he rented a pleasant country house at Dinting Vale, near the point where the River Etherow emerges from the steep-sided valley of Langendale, after its descent from Saltersbrook and Woodhead. He furnished it in a

style suitable to a professional man in his position, and established there
his two youngest children, Henry and Olinthus, in the care of a house-
keeper, a redoubtable Irishwoman called Mrs Turner, who had already
served him in this capacity in Dublin. The establishment was completed
by a young tutor in holy orders, Mr Langhorne, by James the groom, who
had brought up the ponies and carriage from Peckham, and by the usual
quota of servants, pigs and poultry, as befitted a gentleman's household.
The young brothers were joined by Hutton when on holiday from his
French school, and Olinthus recalls (in an unpublished autobiography)
that life at Dinting Vale was generally happy, although the consequences
of boyish misbehaviour when their father was at home were liable to be
painful. Young Charles was still in Germany, while the family had recently been increased by the birth of Vignoles's first grandchild, Charles
Croudace. Father and daughter had become sufficiently reconciled for
the former to pay the expenses of Camilla's confinement.

The old liking for cutting a dash had not decreased with the years.
Early in 1838 he had given a ball in Dublin for his friends and their
families, at which over a hundred guests sat down to supper. Now, at
Dinting Vale, on 12 September he entertained the Directors and officers
of the Company to dinner, at a cost which he estimated at £100, including
wine and carriages, 'being about three times what it would have cost at
a first-rate hotel in London'. A week later he took a rare day's holiday
to make one of a carriage-party with Thomas Ellison, a director, and two
other gentlemen, to see the St Leger at Doncaster Races, arising in the
small hours and returning to Dinting Vale at midnight. The diary does
not record whether any of the party backed a winner.

With the Company's share subscription list complete and the first contract let, the time had come for a formal beginning of the work. As a sym-
bol of the Directors' determination to surmount all difficulties, it was decided that the first sods should be cut at the point where the biggest challenge was to be met, the western end of the Woodhead Tunnel. At this
spot the narrow valley of the Etherow, today the site of three reservoirs,
but at that time the course of a meandering mountain stream, opens out
below a steep and rocky bluff, where river and road make a sharp turn
to the west after a steep descent from the moor over its southern shoulder. Here on 1 October 1838 the new line was inaugurated, in a ceremony
in which the first sod was cut by Lord Wharncliffe, the Chairman, who
declared the ground duly broken for the new railway; the Chief Engineer
then cut the next sod, followed by each of the Directors in turn, by the
assistant engineers, by Lord Wharncliffe's sons, and finally by Henry and
Olinthus Vignoles. According to a contemporary account in the *Manchester Guardian* the numerous ladies present 'witnessed this ceremony
from a neighbouring knoll and appeared much amused with the awkward
delving of some of the Directors and young men'. The day was unusually
sunny and clear, and the white marquees, in which the ladies and gentlemen were to partake of a cold collation (provided by the Chief Engineer),
and the white flags marking the intended course of the railway for miles
along the valley, stood out in contrast with the sunlit heather. While the

gentry enjoyed speeches and champagne, about a hundred navvies were regaled with bread and ale on the grass by the roadside. In replying to the Chairman's toast of the Chief Engineer, Vignoles expressed confidence in his plans and his ability to carry them out. Two points in his speech are significant, in the light of subsequent events: the first his statement that the only question unresolved was the nature of the material through which the tunnel was to be driven; and his declaration that his own professional reputation, and his own fortune even, were staked on this undertaking.

Vignoles records that 'the afternoon passed off as usual on these occasions', and we may imagine the scent of cigars mingling with the tang of heather, and the laughter and shouting of young men and boys racing and playing games. At the end of it all young Henry Vignoles fell in the stream, and was obliged to run behind the four-horse brake on the way home, to dry his dripping clothes.

During the afternoon the Chief Engineer showed off a special boring tool he had obtained with the help of his son from Germany. According to the *Manchester Guardian* account, it consisted of a cruciform iron chisel, weighing about three hundredweight, which was raised and let fall by means of a rope over a pulley, at a rate of 60 strokes a minute. At each stroke the operator rotated the chisel slightly, to achieve the boring effect necessary to drive the tool into the rock. Vignoles claimed that it was capable of drilling out a borehole of 12-in. diameter through solid rock at the rate of 6 ft per day. James Sidebottom and Rhodes were particularly interested in this apparatus, and on 15 October they spent the whole day with Vignoles, in drenching rain, setting it up in a quarry on the moor about a mile from the tunnel entrance.

Bad weather was not to be the only cause of hardship and discomfort to the labourers who toiled on the Sheffield & Manchester line. In his book *The Railway Navvies*, Terry Coleman, referring to the building of the Woodhead Tunnel as 'the most degraded adventure of the railway age', has described the appallingly uncomfortable and dangerous conditions in which the navvies worked and lived on the exposed Pennine moorland. Vignoles was no different from any of his contemporaries in believing in the justice of a system built on the exploitation of cheap labour. But he was a humane man, and as a good army officer he believed in getting the best out of his men by looking after them. From the very first he stressed the importance of providing adequate shelter for the workmen on the tunnel and on the more remote parts of the line. As early as May 1837 he recommended that huts should be built for the men before any work on the tunnel was begun. In June 1838, barely three months before the cutting of the first sod, nothing had been done, and Vignoles again urged the Directors to buy land at either end of the tunnel where cottages and shops could be built for the army of workmen; in the meantime he had used his influence with the Army Ordnance department to procure three hospital marquees and ten bell-tents for use during the summer.

With the coming of winter the Board authorised Vignoles to begin providing permanent accommodation, but although the first borehole had reached a depth of 140 feet by January 14, bad weather, shortage of money and the many calls on his time at other parts of the line delayed any serious work on the tunnel until the following summer.

By the spring of 1839 the staking out of the whole line was complete, and negotiations for the purchase of land were proceeding, the stretch from the crossing of the Etherow at Saltersbrook to Broadbottom being purchased from the Duke of Norfolk at a price advantageous to the Company. In June the 'Gamesley Contract' (no. 5) from Broadbottom Viaduct to Dinting Vale was let to Miller & Blackie of Liverpool. Meanwhile, the Directors had authorised the sinking of four more boreholes and the first shaft on the line of the summit tunnel, and Vignoles was again calling for houses, cottages, stables, stores for tools and machinery and proper roads and approaches across the moors. The Directors approved the contracts for these by the end of May, but the subsequent history of the works shows the provision of accommodation for the navvies to have been grossly inadequate. The supply of buildings was from the start months behind the demand, with the result that the majority of the men were obliged to build their own huts, mostly of stone without mortar, and roughly roofed with thatch or flagstones. In these they lived in conditions of the greatest squalor.

In March 1839 Vignoles had a long discussion of his plans with Sir John Burgoyne, who came over from Ireland on a visit and showed a lively interest in the work. Shortly afterwards (on 19 April) Vignoles presented revised plans for the construction of the tunnel. These were discussed with the Directors at several informal meetings, and the final conclusions are set out in an entry in Vignoles's diary for 1 June. The new plans were

Western end of the Woodhead Tunnel, before the opening of the new tunnel in 1954, on right. The building above the tunnel is the public house built for the navvies, now demolished. On the moor above, one of the observation towers recommended by Vignoles is still standing.

to provide for two single-line tunnels, 12 ft wide and 18 ft high, the centre of each tunnel being at a distance of 16 ft from the centre line as laid out on the ground. This would leave a pier 20 ft wide between the tunnels. Only the southern tunnel was to be constructed in the first instance. Fourteen exploratory boreholes would first be sunk along the centre line between the tunnels, followed by six shafts 9 ft in diameter, from the base of which driftways 4½ ft wide and 5 ft high would be broken out laterally and then driven along the top and bottom of the tunnel. Similar driftways would be driven from each tunnel face. The main tunnel would finally be opened out from the driftways. The length of the tunnel was estimated at 3 miles 13 yards, rising steadily at a gradient of 1 in 200 from west to east, being 857 ft above sea-level at the Woodhead end and 943 ft at Dunford Bridge.

Vignoles recommended that steam pumping-engines should be installed at the top of each shaft, and, in addition to the engine-houses for these, stone observatory towers were to be built at intervals on the moor for the mounting of the surveyors' instruments. A special tunnel sub-committee was appointed to advise and take decisions on the day-to-day working of the tunnel, and a new tunnel engineer, William Cooper, was appointed.

Vignoles's original contract was now coming to an end, though there seemed no doubt that he would be appointed Chief Engineer to complete the work. However at this point a disagreement arose as to the terms of his appointment. Vignoles claimed that he should have full responsibility for all matters connected with the engineering department, including the right to choose and appoint his own assistants. Although Lord Wharncliffe and the Director Michael Ellison, with whom Vignoles was invited to discuss his terms, seemed disposed to agree, the Board as a whole thought otherwise. They proposed that the responsibility of the Engineer-in-Chief should be confined to that of advising generally on engineering questions, without any authority to act independently of the Board, and that the appointment of any resident engineer should be made by the Directors, after due consideration of recommendations made by the Engineer-in-Chief. Vignoles accepted these proposals, with their extremely vague definition of his duties and responsibilities, with the greatest reluctance, and, as he noted in his diary on 24 July, only 'on the repeated assurance that I should not be interfered with. (We shall see how they keep their word – but I shall have grave doubts I shall be obliged to give up before Xmas.)'

The disagreement which drove him to thoughts of resignation was the culmination of differences which had always existed between him and those members of the Board who considered that their engineer was altogether too big for his boots; while on Vignoles's side the interest shown in the work by some of the Directors had become too much of a good thing, developing into downright interference.

As though to show how little he cared for their opinions, he now staged a demonstration of self-confidence on the most lavish scale. His eldest son Charles had recently returned for good from Germany and had been helping the assistants on the line. On 19 August his father gave a magnificent banquet at Dinting Vale to celebrate his coming of age. (In fact Charles was only 20, although entering his twenty-first year.) Over a hundred guests attended, there was dinner and dancing in a marquee on the lawn, and although outdoor amusements and fireworks were spoiled by a continuous deluge of rain, a good time was had by all. While noting that the whole entertainment went off exceedingly well, Vignoles made the wry comment 'but after all to the great body of people it was like *throwing pearls before swine*', coupled with a reference to one of the Directors, old Mr Sidebottom, who drank two bottles of burgundy and 'got himself drunk'. Olinthus records his recollection of pouring rain, and his astonishment at the dexterity and balancing powers of white-waistcoated and white-cravatted waiters, supplied by the proprietor of the Liverpool Angel Hotel. Vignoles recorded that the total cost of the entertainment amounted to £300, which included hospitality (probably liquid) extended to the neighbours, who flocked in crowds, in spite of the rain, to see the fireworks. Young Charles carried through the honours of the evening very well, but his father confessed for once to being 'completely knocked up'. Not long afterwards Charles was sent off to work for a spell as assistant resident engineer on the Midland Counties line.

From 1 June 1839 the Tunnel Committee met monthly, and work gradually made progress. The upper and lower driftways at the western tunnel face were begun, while at the eastern end two shifts of 22 men, each working for 12 hours on end, started digging an open cutting from the tunnel face to Dunford Bridge. On 19 October it was reported that at the western face the upper and lower driftways had penetrated 63 yards and 52 yards respectively in strong grit, while at the eastern end the upper driftway was 8 yards in, contending with strong beds of rock, laminated with thin layers of shale. Five shafts had reached an average depth of 20 yards, where water had already been encountered, and the work was held up, waiting for the delivery of pumping-engines.

Meanwhile, in spite of his agreement to the Directors' terms, Vignoles continued to chafe under the conditions laid down for the conduct of the engineering department. But a far more serious situation was developing over his holding of the Company's shares. It was a time when money was short and many shareholders were failing to respond to the calls made on them in respect of the holdings. Hitherto the Board had been somewhat lax in pressing for payment, but money was now required to push on with the works, and those shareholders who had dutifully paid up were demanding that the Directors should adopt a tougher line. So far no calls had been made on any of the 1400 shares held by Vignoles and his friends; but now a strong body among the Directors demanded that they should be declared liable for the three calls already made, amounting to £10 per share, and for all subsequent calls that might be made. Faced with the possibility of being required to pay £14 000 immediately, with the threat of much higher demands to come, Vignoles replied by demanding payment for three months' extra time he had been obliged to work to fulful his original contract. This, together with money he had spent on the Company's behalf, he calculated at £3000. At the same time he complained of interference by members of the Board in the running of the engineering department, and insisted that the arrangements for the department should be reconsidered as a condition of any discussion of his share liabilities, and of his claims against the Company. The atmosphere in Board meetings became charged with tension. On 11 September Vignoles went so far as to refuse to sit down in the Board room so long as a certain Mr Bennett, with whom he was particularly at odds, was present; and a fortnight later he informed Lord Wharncliffe and the Deputy Chairman that he would resign, unless the arrangements for the engineering department were changed.

This was of course exactly what his enemies on the Board wanted; and it is even probable that, knowing his fiery temperament, they hoped to goad him into doing so. What must have added to his exasperation was the thought that after so many delays some progress was being made on the construction of the tunnel. About this time Mr and Mrs Pim came over from Dublin to stay at Dinting Vale, and Vignoles's pride in showing his Irish colleague over the works must have been tempered by the threatening clouds overhanging his future. Night after night he sat up into the small hours, calculating how he might raise the money he owed. Then, in the middle of October, after a particularly stormy encounter with the Directors, he sought a brief respite from his worries in a four-day carriage tour through Derbyshire with his Irish friends and his three youngest sons. At Chatsworth Mr Paxton showed them the great conservatories being built, at Matlock they admired the scenery from the Heights of Abraham above the Derwent gorge, and at Cromford they 'took a peep at the domain and house of Sir Richard Arkwright'. After walking several miles with the Pims through Dovedale on a 'lovely serene day' Vignoles took them on to Derby and the Trent, where they inspected the recently completed Sawley Bridge. Here they met young Charles and walked over part of the Midland Counties line with him. It was a brief idyllic interlude, from which Vignoles travelled by night coach to London to explore ways of raising money in the City, before returning once more to face the intractable problems awaiting him at Dinting Vale.

He was greeted with a report from Cooper, complaining of further interference in the engineering department. Vignoles immediately ordered Cooper to take no instructions except from him. But not all the Directors were against him; and amongst those who were favourably disposed was the Chairman, Lord Wharncliffe. At a special general meeting of shareholders on 24 October his Lordship referred to the difficulty the Directors were in with regard to the 1402 shares held by one person (without mentioning Vignoles by name). He was fair enough to point out that this gentleman had bought shares only with the good of the Company in mind; and that although he could subsequently have sold some of them at a profit when the shares rose in value, he had refrained from doing so. The shareholders voted to leave matters in the Directors' hands. Vignoles was called to meet the Chairman and a committee of Directors, who agreed after a long discussion that he should forfeit all the shares he had bought or subscribed for, including those assigned to his friends; at the same time he was personally assured by the Chairman that neither he nor his friends would be held liable for the calls hitherto made on the shares. On the other hand the Directors refused to admit his claim for payment for the extra time he had spent in setting out the line and making the working drawings.

Matters having reached this stage, Vignoles gave notice of his intention to resign, as soon as he would be in a position to hand over to a successor. On 15 November, Nicholas Wood having been invited by the Board to look over the railway plans and sections, Vignoles submitted a formal letter of resignation to Lord Wharncliffe, 'after a very distressing interview'. This was accepted at the December meeting of the Board. The Directors passed a resolution expressing regret and stating that they had no cause and therefore no intention to cast the slightest imputation on Vignoles's professional character. The same meeting decided to invite Joseph Locke to fill the vacant post of Engineer-in-Chief.

Vignoles's diary for that day concludes with the note: 'Thus ends my connection with the Sheffield & Manchester Railway after having given great attention to it for nearly 4 years and having sustained in one shape or another an actual loss of £10,000 in hard monies.' It was a bitter enough blow, but he did not then know how much worse his position was ultimately to be. For the moment, now they had got what they wanted, the Directors could afford to make a show of generosity. They appointed a sub-committee to consider Vignoles's hand-over and the settlement of his outstanding claims, and on 18 December voted to pay him £1000 on account of cash disbursements on the Company's behalf. This meeting was attended by Joseph Locke, who agreed to accept the post of Engineer-in-Chief, on the very terms which had been so fiercely disputed between Vignoles and the Board: that he should have complete responsibility for the works, as well as the right to select and appoint his own resident engineers. The Directors could hardly have been more insulting to their former Engineer-in-Chief.

The hand-over was completed by the end of March 1840, by which time Locke and Nicholas Wood had commended the quality of Vignoles's plans and sections to a shareholders' meeting. Substantial progress had been made by the force of 400 navvies toiling on the tunnel works; driftways were slowly advancing from either face, the five vertical shafts were making good progress, surface roads had been laid to the site of each shaft, observatory towers and even a few cottages were slowly making an appearance.

The basic plans for the line, originally agreed between Locke and Vignoles, remained unchanged, but Locke later modified the design of the Etherow and Dinting Vale viaducts. Vignoles records in his diary for 24 April 1839 that he had settled with the Board for a stone viaduct at Dinting Vale, rather than a timber structure on stone piers, which was the plan for the Etherow crossing at Broadbottom. At the same meeting, however, it was decided that Vignoles and a sub-committee of Directors should visit the laminated timber viaducts just completed by John Green at Ouseburn and Willington, on the Newcastle & North Shields Railway. Vignoles had already seen these in course of construction, when he was attending a meeting of the British Association at Newcastle in 1838. As a result of their visit, the Directors opted for the more economical construction in timber rather than stone, and Vignoles planned both viaducts with main arches of timber on stone piers. He favoured a flatter arch than Green's, thereby reducing the initial bending stress in the timber, his central arch at Etherow being designed with a span of 180 ft and a rise of 25 ft, with a 170-ft radius of curvature.

On Nicholas Wood's advice (after consultation with Green), Locke reverted to arches of smaller radius, on the grounds of greater rigidity; for example the Etherow viaduct was ultimately built with a central span of 150 ft and a rise of 40 ft, a change of which Vignoles was later to challenge the necessity. We shall have more to say about Vignoles's views on timber bridges later in this book (see Chapter 12).

As regards the tunnel, Locke followed the plan as revised by Vignoles, only reducing the number of shafts from six to five; but the difficult nature of the beds of shale and millstone grit revealed by the first sections of the tunnel excavations, and the unexpectedly large amount of water encountered, caused him within 18 months to raise the estimate for its construction from £106 000 to £207 000, a sum which approximated very closely to its final cost. The cost in human lives and suffering, among the men who constructed it, was also to be heavy. Thirty-two men were to die, and more than two hundred to be seriously injured, before it was finished in 1845.[1]

Stripped of his holding of shares, and having terminated his contract with the Company, Vignoles might have been forgiven for thinking that his fortune had reached its lowest ebb. But even worse was to come. On 31 March, he was told by Lord Wharncliffe and Michael Ellison that the Board were likely to go back on their promise not to ask for payment of the calls on the shares previously held by him and his friends and recently forfeited. He cast about desperately for a solution, calculating that if all his property were realised, including the sums owing to him from all quarters, between £18 000 and £20 000 might be raised to satisfy the Board's demands. He was also prepared to forgo all the Company owed him. But during May circulars were sent out by the Secretary demanding payment from Vignoles's friends. Vignoles tried vainly to see Lord Wharncliffe. The diary entries become more and more desperate. To add to his distress, his forty-seventh birthday passed with no visit or letter from Camilla. Then on 19 June the bombshell finally burst. Vignoles arrived at Dinting Vale to find a letter from the Chairman, stating that the Board had formally refused to recognise the pledges given by him and the Deputy Chairman to Vignoles in October, and had resolved to sue for payment for all the calls due, including two further calls made since the October meeting. This additional claim, outrageous as it might seem to be, had a technically legal basis, in that the forfeiture of the shares had been a condition of the Chairman's pledge to Vignoles, and that with the pledge annulled it only became effective from the Board's latest meeting. The result was a demand for a total of £25 per share, amounting in all to about £35 000. The majority in favour of this decision was only a narrow one, and in consequence of it Lord Wharncliffe, Michael Ellison and several of the Sheffield Directors had resigned.

While desperately anxious about his own position, Vignoles was chiefly concerned as to what could be done to protect his friends, many of whom were quite unable to pay the sum demanded of them. His first action was to make over his whole property to four trustees, empowered to treat with the Company on his behalf and to give financial help to his friends. On counsel's advice an injunction was applied for, to restrain the Company from acting. Meanwhile Vignoles considered how he was to dispose of his property and at the same time make a home for his family. Horses and carriages, pigs and poultry, were sold; the lease of Dinting Vale House was made over to Alfred Gee, Locke's new resident engineer. Fortunately Charles and Hutton were provided for, the first having just been appointed assistant engineer to the Shannon Navigation Commission, the latter apprenticed to William Fairbairn in Manchester. But the two youngest boys would shortly be home from Repton for the holidays, the last they were to spend at Dinting Vale. On 14 September, Vignoles took the lease (at a rent of £20 per annum, including rates and taxes) of a small house in Manchester, retaining the faithful Mrs Turner to make a home for Hutton and the two younger boys, who would leave Repton and go to Manchester Grammar School in January.

On 7 September, all attempts by Vignoles's trustees to persuade the Directors to compromise having failed and the injunction having been refused, the case against 18 shareholders was heard in Liverpool. The plaintiffs did not have it all their own way. Counsel for the defence pointed out that Vignoles had acted in good faith and with the full knowledge and agreement of the Board, and argued that since the Board had failed to make calls when they were due on any of the shares held by the defendants, the latter had a right to assume that they must be exempt from payment. In the course of the proceedings considerable irregularities in the administration of the Company were exposed; but in spite of this, in the end judgment was given against all 18 defendants, though counsel were allowed to appeal on various points of law. Vignoles wrote 'the base Conduct of the Directors was exposed in a very complete manner and produced a great effect in Court', while the case was the subject of much adverse comment in the Press.

On 29 September, he spent his last night in the country home from which he had directed what had promised to be the greatest operation of his engineering career. In view of the part Russia was to play in the recovery of his fortunes, it is interesting to note that he had as his guests the Russian general Tcheffkine and his A.D.C., to whom he was acting as guide on an extensive tour of British engineering projects. For despite all that had happened, his busy professional life went on, and he maintained a bold face to the world; only the periodical entries in his diary reveal the anxieties he was suffering while awaiting the results of the appeal to the Court of the Exchequer.

It was not until 15 January 1841 that the Court's decision was made known. The appeal was disallowed on all counts. Vignoles was in despair. He wrote on the 16th 'the judgment having gone against my friends considering in what mode I could assist them to meet the cruel pressure that would immediately be put upon all of them – half distracted at the frightful prospect before us all . . . Good God! that men whom I have served so faithfully & for whose Railway I had done so much should act like this.' For some time he confessed to being quite unfit for business: 'The dreadful burden weighs me down beyond all sufferance and I shall certainly sink under it.' It was a burden he was to carry for some years. Olinthus Vignoles estimates his father's total loss at £80 000. That Vignoles did not in fact sink under the burden, but continued to struggle until he had paid his debts and rebuilt an apparently shattered career, is a measure of the extraordinary resilience of his character.

8 The difficult years, 1841–1844

As Vignoles surveyed the ruins of his career in the early months of 1841, the pages of his diary reiterate not merely his fears for his own future, but also the remorse and anxiety he felt on behalf of his friends. The Directors of the Sheffield & Manchester Railway were showing them no mercy. Vignoles was deeply shocked to hear that two of them had been committed to prison for debt, and sought desperately for a means of relieving them. On 29 May he wrote: 'such a mass of misery and distress [is] thrown on my friends that I think I shall go mad.' And it was only in July that he began to dare to hope that he himself would escape the ultimate disgrace of going 'through the gazette' (i.e. bankrupt).

The loss of position and prestige entailed by his resignation he does not mention; but one can hardly doubt that he felt this deeply. It was a blow to his pride; and he required not only money but position to recover from it. In the search for this he was to pass several frustrating years. About this time his diary, besides being a record of his daily work, becomes more and more the repository of his thoughts and feelings; and the entries emphasise the extent to which he was facing the struggle alone.

His aunt Isabella Hutton, with whom he had enjoyed a very special relationship, had died in 1839. He felt her loss deeply. Ever since his rejection by his grandfather, the need to justify himself in the eyes of his family and the world had been a driving force in his character. And when, in his married life, admiration and respect had been finally supplanted by nagging criticism, he turned to Miss Hutton for what he no longer found in his wife. Not that the aunt was not often critical of the nephew; but because she was his aunt, he would put up with criticism from her which he would not have suffered from Mary. With Miss Hutton's death, intimate family relationships were only possible with the younger generation. Unfortunately, like many a Victorian parent, Vignoles found it difficult to step down from the father's position of authority, however much he might wish to enjoy his children's sympathy and enter into their lives.

His first concern now was to continue to make a living, out of which something could be saved to clear him of the burden of debt. Several of the other projects he had on hand during his work on the Sheffield & Manchester Railway were coming to an end, and in his search for new work he seemed to be dogged by bad luck. A proposal for a London & Chatham line, under the chairmanship of Isaac Goldsmid, son of the agent for the S.&M.R., came to grief, leaving Vignoles with a further liability for money spent on preparing the plans; a new proposal for Irish railways, brought forward by Lord Morpeth in the House of Commons, was unsuccessful; and possible openings in France and Russia were also disappointing. In 1839 he had been engaged as a consultant to report on the Paris–Versailles line, and he had kept in touch with M. Léo, its managing director. Now that Locke and Brassey, with an army of English navvies, were building the Paris to Rouen and Le Havre line, Léo proposed a shorter route by way of Calais or Boulogne. He exchanged long letters on the subject with Vignoles, who interested Robert Stephenson and some of the Directors of the South Eastern Railway. But the City showed no enthusiasm for ventures across the Channel. Correspondence with Colonel Du Plat, British Consul in Warsaw, about a possible line from Warsaw to Cracow and St Petersburg, proved equally unfruitful.

Among all these disappointments, two minor projects brought Vignoles some relief. One was the construction of a floating steamer landing-stage at Southwark Bridge, a comparatively simple problem, but one to which he gave much skill and attention. A rather more complex affair was the designing of a patent slip-way for the West Indian island of St Thomas. Its construction took about two years, the slip-way being built in sections in Glasgow and shipped out to St Thomas (to be assembled *in situ*) together with the machinery and even the stone for the quay and foundations. The project was a comparatively lucrative one, although it was to be some time before he was able to reap any financial benefit from it.

It was also in this very worst period of his fortunes that he first became interested in a new and remarkable development in railway propulsion. During 1840 and 1841 the brothers Jacob and Joseph Samuda and their partner Samuel Clegg were carrying out trials of their so-called 'Patent Atmospheric Railway', on a stretch of the partly-built West London line at Wormwood Scrubs. The atmospheric system of propulsion had been developed as an alternative to steam by several different inventors. In the form invented by Clegg and the Samudas, it consisted of a train of carriages propelled by atmospheric pressure acting on a piston moving in a tube laid between the rails. Stationary pumping-engines, placed at intervals along the line, exhausted the air from the tube, with the result that the piston was sucked along it, drawing the train after it by means of a lever attached to the leading carriage, running along a slit in the top of the tube. A leather and metal flange opened to allow the passage of the lever, and was closed after it had passed by a device on the train. While there were plenty of mechanical difficulties still to be overcome, the Wormwood Scrubs experiments had shown that the system could be made to work, though so far no railway company had ventured to adopt it.

Vignoles had at first been sceptical about the system. He had witnessed one of the Wormwood Scrubs trials (accompanied by M. Léo who was on a visit to England) on 11 June 1840, when he wrote: 'My first view fully confirmed as to its practicability; but suppose it to succeed to the uttermost it *will* only be a system of stationary engines substituting a Tube for a Rope.' But before long he became one of the system's most staunch supporters. Its novelty and ingenious simplicity appealed to his romantic imagination, while with his usual optimism he was prepared to brush aside its technical defects; and during the next few years, until the atmospheric railway eventually expired as a victim of its own inherent weaknesses, he was continually recommending its adoption. Among those he interested in the system was his old friend James Pim, treasurer of the Dublin & Kingstown Railway. Pim was trying to revive the project of an extension of the D.&K.R. as far as Dalkey, following the line of a tramroad used by the Kingstown Harbour Works Commission to bring down granite from the Dalkey quarries. One of the advantages of atmospheric propulsion was that it did not depend on the friction exerted between the locomotive driving-wheels and the rails. Consequently it was suitable for use on quite steep gradients, a circumstance which Pim seized on as a recommendation for its use not merely on the Dalkey line, but in the wider development of Irish railways, in which both he and Vignoles were deeply interested, following on the report of the Irish Railway Commission. With a view to enlisting the aid of public opinion, Vignoles collaborated with Pim and Mahony in the preparation and publication of two pamphlets, in the form of open letters issued over Pim's signature, one addressed to Lord Morpeth, Secretary of State for Ireland, the other to Lord Ripon, President of the Board of Trade. Pim claimed that a system of light railways, such as the atmospheric railway could supply, demanding a minimum of earthworks and built at low cost, was particularly suited to Ireland.

The outcome of this direct approach was a Government investigation and report, by Lt. Col. Sir R. Smith, R.E., and Professor Peter Barlow, published in March 1842. This concluded that the atmospheric principle was established, but that the cost of installing the tube and the stationary engines was likely to outweigh any savings there might be on construction; and it pointed to many unsolved practical difficulties in the system 'in regard to junctions, crossings, sidings and stoppages at road stations'. However this did not deter the Directors of the Dublin & Kingstown Railway from deciding to go ahead with the Dalkey extension, for which the Irish Board of Public Works was ready to supply a loan of £25 000. Vignoles was appointed as engineer, with William Dargan as contractor, and the Samuda brothers undertook to supply and install the atmospheric apparatus.

As the tramroad was already in existence, no elaborate engineering work was required, and to judge from his diary Vignoles must have left most of the work to Dargan and his resident engineer. Vignoles himself continued to take an active part in the public debate inspired by the Barlow report and the continuing experiments at Wormwood Scrubs. He sent copies of the report, and of Pim's letters, to his friends in France, Germany and Poland; he gave explanations to visitors at the trials, and in June 1842 he delivered a lecture on the system to a meeting of the British Association in Manchester, illustrated by models, drawings and tables of calculations. In this he ignored the findings of the Barlow report as regards the cost of installation, maintaining that the removal of the locomotive and tender must reduce the average load by half, and that this reduction in weight, extending beyond the rolling-stock to the works of the line itself, would result in a considerable saving of expense. He also claimed that as the locomotive was a frequent cause of accidents, the atmospheric system would make for greater safety, especially when only a single line was used. Of the other difficulties instanced in the report he made no mention.

While the construction of the Dalkey extension proceeded, the opinion of engineers remained divided as to the atmospheric system's merits. Brunel, Vignoles and Cubitt were among its chief supporters, Locke and the Stephensons were against it. In August 1843 the first trials on the Dalkey line were held; on the 19th a train reached an uphill speed of 28 m.p.h.; Joseph Samuda himself rode on the piston carriage on this occasion, and the Lord Mayor of Dublin was among the distinguished passengers. During the autumn further trials took place; among the many visitors were I. K. Brunel, who was shortly to introduce the system on his South Devon Railway, Major-General Pasley, Inspector-General of Railways, who gave it a favourable report, and M. Mallet, Inspector-General of Public Works in France, who returned home to recommend that his government should give the system a trial. Meanwhile Vignoles, Pim and Samuda examined the possibility of using the towpath of the Thames & Medway Canal for an atmospheric line, from Gravesend to Rochester by way of the canal tunnel; Vignoles considered using it for a line from Tonbridge to Tunbridge Wells; and Pim canvassed support for a similar line westward from Dublin along the banks of the Grand Canal, at that time the main artery of transport from east to west in Ireland.

The Kingstown & Dalkey line was opened to the public in March 1844. It was 1¾ miles long, and almost entirely uphill to Dalkey at a gradient of 1 in 115, with a 440-yd stretch of 1 in 57 at the Dalkey end. Trains made the uphill journey to Dalkey by atmospheric power, and returned to Kingstown by gravity. Pim and Vignoles had every reason to be pleased with the railway's performance. Once it was in operation, a half-hourly service of trains was maintained throughout the day, at an average speed of just over 26 m.p.h. for the outward journey and 18 m.p.h. for the return. In the first 11 months of service, the line worked on 337 days. 17 506 trains were run, averaging 17 tons in weight, and made up of two to five coaches, each carrying an average of four passengers. During tests the maximum speed reached was 51½ m.p.h., which considering the train had to negotiate three curves in the line of 570-ft to 700-ft radius was perhaps fast enough. There were however two occasions when the piston carriage escaped on its own owing to the brake being released accidentally, when speeds of 60 to 80 m.p.h. were said to have been reached.

Power for the line was supplied by a single engine, built by William Fairbairn of Manchester. It had an enormous 36-ft diameter fly-wheel, and drove an air pump with a cylinder 66½ in. in diameter, with a 66-in. stroke. This was more than sufficient to create the necessary vacuum in the 15-in. pipe laid between the rails. The working of the line was complicated by the fact that the engine-house was at Dalkey, whereas the trains requiring power started from Kingstown. Eventually the newly invented Wheatstone telegraph would ensure communication between the engine-man at Dalkey and Kingstown Station. But in the early days the engine-man had to rely on the time-table, and start pumping just before the train was due to leave, hoping that all was well at the other end of the line. About 45 seconds of pumping was sufficient to start a train of medium weight, after which the brakesman, if he had his eye on the pressure gauge in his carriage, would release the brake and the train would move off; pumping continued during the running-time of four minutes, steam being shut off when the train appeared in sight of the engine-house.

Among the many weaknesses of the system, not the least of which was the difficulty of ensuring and maintaining an efficient seal on top of the atmospheric tube, this fact that it was the man in the engine-house, rather than the driver on the train, who was in control of the motive power was one of the most important. Considering all things the Kingstown & Dalkey Railway was remarkably successful; but it had the advantage that it was a short single-line shuttle service, with only one engine-house, and it only required power for trains going in one direction. For a longer line, where a succession of engines was needed at two- or three-mile intervals, the difficulties were multiplied, not only as regards the passage from one source of power to another, but also where points, crossing-places and sidings at stations necessitated complicated valves and joints on the atmospheric tube. It was these difficulties in particular which were pointed out by Robert Stephenson in a report made by him to the Chester & Holyhead directors soon after the K.&D.R. opened. While admitting the advantages of the atmospheric system for short lines, he maintained that for long lines with heavy traffic the system would be far too inflexible, seeing that it would only need the failure of one pumping-engine for the whole line to be brought to a standstill. Even so ingenious a mind as that of I. K. Brunel was defeated in the end by the various practical difficulties of the system, though his South Devon line ran for 16 months, as did Cubitt's London & Croydon line. In comparison the K.&D.R. maintained a regular service for ten years, so that Vignoles enjoyed the double distinction of being engineer not only to the first commercial atmospheric line, but to one which outlasted all its British rivals. Its length of service was only surpassed by that of the Paris–St Germain line, which operated from 1847 to 1860, having the same advantages as regards gradient and density of traffic as those enjoyed by the K.&D.R.[1]

The search for employment led Vignoles into a new field when in June 1841 he accepted the post of Professor of Civil Engineering at University College, London. It was a two-year appointment, but the subject attracted few pupils; and though his introductory lecture, delivered on 10 November 1841, was attended by about a hundred people, only a dozen of these were regular students. Vignoles records that he spoke from notes and that the lecture was 'received with the most profound attention'. He took great pains in the preparation of his lectures, enlisting the help of his assistants in the provision of models and diagrams to illustrate them. With a characteristic eye to the audience beyond the academic world, he was equally careful to check the final text with the shorthand writers who came to report for such papers as the *Railway Times* and the *Mining Journal*. According to Olinthus Vignoles his father intended to use the course as the basis of a treatise on civil engineering, but he never had sufficient leisure to settle down and write it.

From what we know of Vignoles's relationship with his assistants it seems likely that he was more successful as a practical teacher than as a lecturer; for instance, he took his pupils to see the demonstrations at Wormwood Scrubs, and to study various construction sites, such as the scene of a landslip on the Croydon & Brighton line at New Cross, where he pointed out the effects of insufficient care in the construction of earthworks and retaining walls. He always preferred to demonstrate his conclusions in the round, and 'on the ground' if possible. Hence his habit of using models as well as drawings to illustrate his work, and his delight in conducting visitors to engineering sites all over the country. He was a born showman in fact, and as he was also an excellent linguist he was much in demand as a guide to visitors from abroad. It is clear from his diary that he enjoyed these trips as much as his guests did – particularly when they provided a ride on a new locomotive, such as the American Norris engine toiling up the 1-in-37 Lickey Bank near Bromsgrove, or George Rennie's new *Satellite* locomotive puffing down the London to Brighton line.[2]

Though his appointment at University College cannot have brought him much in the way of fees, it did give him a position in the academic world, on a par with what he had already established for himself in such bodies as the British Association. In 1837 at the Association meeting at Liverpool he had been elected Secretary of the Mechanical Section, and displayed models of his bridge over the Ribble at Preston. In 1840, in Glasgow, he read papers on timber bridges and railway curves and gradients, while in 1842 at Manchester he was to lecture on railway cuttings and on the atmospheric railway, as we have already noted. His interest in the atmospheric system took him in the same year to Cornwall, when he travelled overnight by the G.W.R. to Taunton, and thence by mail coach all day to Falmouth, where he was to lecture to the Polytechnic Society of Cornwall, at the request of a Mr Jordan of the Museum of Oeconomic (*sic*) Geology. It seems that the scientific and mining interests in the Duchy had an idea that Cornwall, like Ireland, might be particularly suitable for the development of the new system, though there is no evidence that any line was in fact planned.

Such activities did a good deal to restore Vignoles's self-esteem; but they did little to fill his purse; and despite the projects to which we have referred, the collapse of the plans for the London & Chatham Railway, following on his disappointment over the Irish lines, intensified his depression over his financial condition. Unfortunately for him the year 1842 brought the climax of the period of industrial and trade stagnation which had continued with varying intensity in England since 1837. 'Nothing doing', 'very dull', 'nothing seeming to stir', are typical entries in his diary. The carrying on of daily business, as well as the advance payment of costs and expenses arising from such work as he had, required continued recourse to loans from money-lenders and his more faithful friends. In October he was in great difficulties getting extensions to a number of substantial loans, including a hard bargain with a Bermondsey hatter for an advance of £1200; and he was only able to meet a bill of £1000 for materials ordered for the St Thomas slip-way by extending a loan he had previously had from Braithwaite.

The household in Manchester continually needed ready money. Although he urged stringent economy on Mrs Turner, the family did not often complain of neglect. The boys wrote cheerfully and were not afraid of asking permission to buy what they needed. Their wants were simple enough: a pair of warm trousers for Hutton for rough winter wear in the workshop, good drawing pencils for Henry, a book for Olinthus. Vignoles seems to have done his best to keep Mrs Turner supplied, and she too wrote freely, saying exactly what she thought in an illiterate but vastly expressive style, retailing all the family news, haphazardly interspersed with accounts of housekeeping items paid for or (more frequently) owing, such as grocer, beer and wine, shoemaker, butcher, poor rate, servants' wages, and sittings in church. Vignoles stayed at the house whenever he was in Lancashire, and joined the family for a couple of days at the seaside at Lytham St Annes, passing the time 'rambling about', or 'playing at cricket' with the boys.

Meanwhile the struggle to pay off his debts continued. Early in 1843, he was hopeful about the progress he had made in meeting the demands of the Sheffield & Manchester Railway, though this had only been achieved with the help of other loans of up to £4000, falling due to be repaid at the end of the year. Meanwhile immediate prospects were again bleak. His contract at University College had ended for lack of pupils, a new possibility in France (a proposal for a Tours to Orléans railway) had fallen through, and in England a scheme for a patent gas-meter, which he had been trying for some time to launch in partnership with Samuel Clegg, showed small promise of returns. His only source of income during the year was likely to be from fees for giving evidence at Parliamentary committees.

However when, in April 1843, he was apprehensive of a complete breakdown, relief came in the shape of an offer from the King of Württemberg. Vignoles's contacts with Germany had been maintained. He had recently discussed with the Duke of Cambridge the possibility of using the atmospheric system in Hanover, and with the help of Pim and Samuda had drafted a memorandum on the subject. He had been deeply shocked by the fire which destroyed much of the city of Hamburg in May 1842, and through his friend Hubbé had offered his services free to the Hamburg authorities for engineering work needed to repair the damage. He was now invited to report on plans for railways in Württemberg prepared by a German engineer, to modify them as necessary and then to carry out his own designs as Engineer-in-Chief. Here was what promised to be a scheme of some importance, in which Vignoles, with his usual visionary eye for possibilities, envisaged the beginnings of a through railway system from London to the Mediterranean. And almost at the same moment, on his fiftieth birthday, the long-hoped-for settlement with the Sheffield & Manchester Railway was achieved. The entry in his diary speaks for itself:

. . . complete this day my 50th Year. Looking back at the twenty years that have elapsed since I returned to England from America . . . I find myself a poorer Man than at my start – and yet on reflection the fault is my own! I have gained Money but have never had the art of keeping it. I have brought up my family in a proper Manner . . . I have established myself a first-rate reputation as an Engineer – & particularly so for Railways. I hope justly. Yet I am overwhelmed with Debts and Difficulties . . . I am dunned for a few hundreds for petty debts, and have Bills falling due & no means to meet them! And yet there is a bright prospect for my future . . . This day I got finally settled with the Sheffd. & Mancr. Railway Co. & got a handsome professional Certificate. My engagement with the King of Württemberg to execute his Railways after having designed them is certain – and with activity, prudence oeconomy & perseverence I shall . . . be able ultimately to meet all my Obligations repay all the losses of my friends, set up my sons fairly in the world and increase my professional reputation and connections!

But before he left for Württemberg there was urgent work to be done in North Wales. This resulted from a move by James Pim, and others with Irish and Welsh interests, to promote their own route from London to Dublin through North Wales before Robert Stephenson's Chester to Holyhead line obtained Parliamentary approval. Vignoles was called upon to make a survey, going once more over the country he had covered for the Irish Railway Commission. He enlisted the support of I. K. Brunel, who promised that the Great Western Railway Company would be interested and would pay for the survey. With a group of six assistants, headed by Purdon and Collister, Vignoles travelled by steamer from Liverpool to Caernarvon, where he hired a four-horse omnibus to convey the team and their instruments to Tremadoc. They worked furiously all through August, reviewing both the coastal and inland routes. In view of the problem which Robert Stephenson was shortly to tackle in bridging the Menai Straits, it is interesting to note that Vignoles prepared two alternative plans for a crossing, one with a four-arched, the other with a seven-arched timber bridge on stone piers.

Leaving his assistants to complete the final details of the survey, he travelled back to London at the beginning of September. At Wolverhampton he fell into the engine-pit while crossing the line in the dark, sprained his foot badly and narrowly escaped being run over by the engine. In spite of this mishap he completed a report to Brunel and Pim, before calling on the Württemberg Consul to receive £50 travelling expenses. So short was he of ready money that without this it would have been impossible for him to set out for Germany. He had even been obliged to write for a loan from his son, Charles, who sent him £20 from

Ireland. And it seemed that the bailiffs were on his trail; for on 16 August he wrote: 'in consequence . . . of report of certain enquiries at Chambers I am very uneasy as to what will turn up in the next fortnight and I wish I was well out of England'.

He set off in high hopes. Camilla and her husband travelled with him for company on the journey, and Hutton, now 19 years old, accompanied his father in his first post as assistant engineer. Vignoles did not allow the considerable pain he was suffering from his sprained foot to spoil his enjoyment of the sights on the way to Stuttgart. But in spite of the rosy prospects which the Württemberg post seemed to offer, he was once more to be cruelly disappointed. He was warmly received by the King and

Hutton Vignoles, aged 19 years. A water-colour painting by C. Müller, dated 1843.

made welcome by the Court. But it soon became clear that neither the Government nor the government engineers looked with favour on their King's idea of introducing a foreigner to be the Engineer-in-Chief of their railways. Vignoles found himself surrounded by an atmosphere of intrigue. Every possible obtacle seemed to be put in his way; he had great difficulty in obtaining essential information, he was given no drawing-office facilities, and he could not trust the translators.

Nevertheless he and Hutton, his only assistant, worked hard through the winter in all weathers, and succeeded in producing a final plan. In this he was encouraged by the continued friendliness of the royal family and the Court, in contrast with the increasingly open hostility of the government officials.

Vignoles's plan included some important architectural features: a city station in Stuttgart built to harmonise with the royal palace and the square in front of it, a decorative colonnade to conceal the railway as it passed in front of the cavalry barracks (cf. Lord Cloncurry's pavilions!), and a private station for the royal family alongside their palace of the Rosenschloss near Cannstatt, a suburb of the city. He strongly recommended the adoption of the atmospheric system for this urban section of the line, in view of possible heavy traffic and the gradient of 1 in 115 running down to the River Neckar; this he proposed to bridge with a timber viaduct on stone piers, consisting of five arches of 160-ft span.

In the end, to Vignoles's great disgust, the King was prevailed upon to reject his plan in favour of one prepared by his own engineers. Bitterly hurt by what he regarded as the King's betrayal of him, he wrote in his diary: 'It is now evident that all my Dream about making Railways in Germany is over . . . the old fool of a King has not the courage to fight for the person he brought here to tell him the truth.'

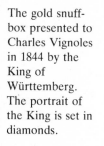

The gold snuff-box presented to Charles Vignoles in 1844 by the King of Württemberg. The portrait of the King is set in diamonds.

No doubt his feelings were soothed by the King's gift of a magnificent gold snuff-box set with diamonds, and by the payment of his fees and expenses.[3] But he was glad to shake the dust of Stuttgart from his feet, and to set out for home – the more so as the Kingstown & Dalkey line had just been opened, and there was news of a more favourable industrial outlook in England.

9 The 'Railway Mania', 1844 –1846

When Vignoles returned from Württemberg in April 1844 he had been away from England for seven months. Letters from his family while he was in Germany show concern for his welfare, and a keen understanding of his situation. Thus Camilla warned him that 'these foreigners . . . are very adept at extracting knowledge from others to serve their own purposes', while young Charles, writing to Hutton from Ireland, where he was working as assistant engineer to the Shannon Navigation Commission, made shrewd comments on the methods of German engineers based on his experience in Hanover: 'they have no knowledge of the practical manner of making preliminary surveys of a Line of Railway . . . with the Eye or at least with a few running levels . . . they go plodding on ahead with a Level and a chain – and know no more than the man in the moon where they are going to', a statement which shows how clearly he appreciated the difficulties his father was experiencing; and he added: 'God knows it makes me so anxious thinking about him that I can hardly give my mind to my own business.'

The remark about Vignoles's methods of surveying deserves comment. He had an eye for country, which was an unerring guide to him in making the preliminary survey of a line. In this he resembled his contemporary Brunel. If we compare the entries in Brunel's diary during his first survey of the Thames valley for the Great Western Railway in 1833, with those of Vignoles when going over the same ground in the following year, the resemblance is striking. We have seen how swiftly the latter covered the country when making the southern survey for the Irish Railway Commission. Yet once a line was chosen he insisted on very careful and thorough calculation of levels and sections by his assistants; and, while checking their findings himself, he would be ready to consider any modifications they might have to suggest. It was of course this delegation of responsibility which made it possible for him to take on so many projects simultaneously, a practice which was to become more frequent with the years, but which was not without its dangers, as we have already noticed.

The interest shown by his children must have been some consolation in his anxieties, as well as the news at the end of January of the birth of his first granddaughter Camilla Mary (destined to become a leading figure in women's education).[1] On a rather different note was the appeal from Mrs Turner, accompanied by a long list of unpaid bills, which arrived just before he left Stuttgart: 'I think you just have forgot that there is five mouths to fill in Manchester and i am sorry to say nothing coming in but what comes from you. . . '

But with the significant change in the industrial climate, it seemed that the worst of Vignoles's financial problems could well be over. The country was entering a boom period, marked by an enormous expansion of interest in railways. This was to culminate in 1845 in the wave of speculative activity known as the 'Railway Mania'. The government was at last beginning to take notice. In 1844 an Act was passed for the regulation of railways, which is chiefly remembered today for having introduced the so-called 'Parliamentary Trains', in a clause which required every railway company to run at least one third-class train per day each way at 1d a mile, for the benefit of the working classes. The following year an advisory board under the Board of Trade was set up, to examine and report on every railway scheme before it was submitted to Parliament. Although inclined to be conservative in approach, favouring extensions of old lines rather than the construction of new ones, the board would have done good work had it survived. But popular feeling against governmental interference was too much for it, with the result that it lapsed in July 1845 – by which time the 'Railway Mania' was in full swing. In any case the board was finding it almost impossible to handle the enormous number of new schemes being put forward, for in the spring of 1845 nearly 300 new lines were applying for Parliamentary sanction.

Here was Vignoles's opportunity. Within three months of his return from Württemberg he was deeply involved in consultations and new surveys in northern and southern England, as well as in Ireland. Their number was to increase during the next three years. Many of the lines concerned remained in the planning stage; others got their bills as far as Parliament but no further; only a few reached the stage of actual construction; but the volume of work was prodigious, and for Vignoles the years 1844 to 1846 were to be as intensely packed with mental and physical effort as any period of his career.

The story of those years, which in many ways epitomises Vignoles's whole working life, is one of countless meetings, conferences and discussions; of long journeys, by coach, private carriage or such railways as were already established; of crossing to Ireland and back; of shorter journeys 'running about' by cab from his office in Trafalgar Square to visit legal chambers and city offices; of hard days across country on foot or on horseback, and long nights at the drawing-board; of hours of attendance on Parliamentary committees, hours of keen attention between gruelling periods of examination and cross-examination; of directing, checking and supervising the work of assistants and pupils; above all, of writing – writing letters, reports and briefs, and the daily record of work in the diary. Through it all he had to preserve the quickness and clarity of mind needed to switch attention rapidly from one project to another, or to struggle with several pressing problems simultaneously.

The railway boom gave particular impetus to supporters of the atmospheric railway system, and naturally Vignoles's name occurs frequently in this connection. In England he was appointed engineer to a projected

line from Dorking to Arundel and Brighton, and to the Windsor, Slough
& Staines Atmospheric Railway. Here he had the satisfaction of prepar-
ing models and drawings of the line for the inspection of Queen Victoria.
In Ireland he was appointed engineer to Pim's Grand Canal Atmospheric
Railway, and of a further extension to the Dalkey line as far as Bray.
While none of these lines came to fruition, Vignoles gave useful support
to Brunel and Cubitt in the hearings of the South Devon, and London
& Croydon bills; and he was one of the principal supporting witnesses
to the Parliamentary Select Committee on the Atmospheric Railway in
April 1845. His advice was also sought from Austria in connection with
a possible line from Vienna to Schönbrunn, and from France on the final
section of the Paris–St Germain Railway. Both these lines, like the Kingstown & Dalkey, and Vignoles's projected line from Stuttgart to
Cannstatt, were suburban railways with steep gradients, and thus particularly suited to the atmospheric principle. Vignoles advised strongly in
favour of the system in each case but, as we have seen, only the French
line became operational.

One of the first projects on which Vignoles embarked in 1844 was a
revival of his scheme for a London to Chatham line, combining with it
a plan to extend the line as far as Canterbury and the cross-Channel port
of Dover. Despite the fact that Vignoles believed he had been let down
by the withdrawal of Sir Isaac Goldsmid's support for his earlier proposal, he does not seem to have objected to Goldsmid's becoming company
chairman once again. Vignoles himself was the prime mover in the
scheme, not merely in the preparation of new surveys but in arousing interest in the City and the Press, and in the various towns on the route,
where he addressed a number of public meetings. The new London,
Chatham & North Kent Railway had to meet the opposition of the South
Eastern Railway Company, which had just completed its own line to
Dover by the rather roundabout route through Reigate, Tonbridge,
Ashford and Folkestone, and now, scenting the danger of competition
from a more direct line, had produced its own plan for a North Kent line
to Rochester and beyond, with Robert Stephenson and George Bidder
as engineers.

The struggle between the two companies, typical of the keen rivalry
which was a feature of the 'Railway Mania', was to last for over 18
months. For a time, indeed, it looked as if they might join forces, an idea
which was warmly supported at a public meeting in Dover, at which Vignoles records that he 'took a very prominent part in the discussion . . . by
a Speech which was very favourably received'. Relations between him
and Stephenson were clearly cordial, and reached the point where
Stephenson made Vignoles some kind of offer. The chairman, however,
was not disposed to treat, and in the end the companies went their separate ways.

In order to cross the Medway at Rochester it was necessary to come
to terms not only with the Admiralty, but also with the Wardens of the
Rochester Bridge Trust. This body had recently decided that the many-
arched bridge it had maintained for centuries must be replaced, and had

engaged William Cubitt to design a new one. However, in answer to the
railway companies' proposals, and in deference to the Admiralty's opinion that a second bridge would prove too great an obstacle to navigation,
the Wardens were now prepared to abandon their bridge and to go halves
in the cost of construction of a joint road and rail bridge with whichever
company was to build the line.

Formal proposals from both companies were received at a special
meeting of the Bridge Wardens on 22 February, when the Wardens stated
their terms, and, without making any binding engagement, expressed
their preference for Vignoles's design of a road and rail bridge of three
cast-iron 167-ft span arches on stone piers, with a swing-bridge section
for the passage of ships. Vignoles records that the two deputations dined
together after the meeting, and that he and Stephenson posted back to
London together. It is not surprising that a few days later the two engineers were again discussing the possibility of a compromise.

Planning and negotiations continued during the summer of 1845, while
the defeat of both Parliamentary bills only strengthened the resolve of
the two companies to resume the struggle with revised plans and new ap-
plications to Parliament. Meanwhile the Bridge Wardens called for new
proposals from both companies, which were put forward and discussed
at a series of meetings between the Wardens and the company representatives. At the last of these meetings, on 3 February 1846, the Wardens
opted for a five-arch stone bridge designed by Stephenson, with a swing
bridge at its western end. The South Eastern Railway declared themselves ready to build this for £50 000. Vignoles, who was representing the
N.K.R., asked for time to consider the matter, noting in his diary that
he feared the Wardens were playing one company off against the other.

The Wardens had still to obtain Admiralty approval for Stephenson's
bridge. Vignoles may well have suspected that their Lordships might object, for the day after the Rochester meeting he took his own plans to
Captain Beaufort, with a request that their suitability might be examined.
When a month later the Bridge Wardens were informed that Stephenson's five-arch design was unacceptable to the Admiralty on navigational
grounds, they also learned that the Admiralty had approved the N.K.R.
design. Despite protests from the S.E.R. deputation, the Wardens now
resolved to adopt Vignoles's design, and invited tenders from both companies for its construction. In the meantime they had covered themselves
against the possible failure of both companies by applying for a Parliamentary bill for the construction of their own new bridge.

In this they acted wisely. Vignoles was once more to be bitterly disappointed. Despite the superhuman efforts he made to ensure that counsel
and witnesses in the Parliamentary committee were fully briefed, and
that the latter, who included Brunel, Locke and Rastrick, had first-hand
knowledge of the proposed line,[2] the North Kent Bill was negatived in
the Commons while the South Eastern were allowed their line, but only
as far as the Medway. Vignoles was sure they could still be defeated by
a petition brought against them by the North Kent, based on the opposition of Goldsmid and other landowners, whose property would be af-

fected by the line. But on 24 July 1846, Goldsmid withdrew his opposition, and the South Eastern Railway Bill was finally approved by the Lords. The decision was endorsed by the North Kent shareholders, on learning that the South Eastern were to pay £30 000 for their freedom. Vignoles, who believed that the victory had been all but won, described the arrangement as 'a most disgraceful transaction', although it at least assured him of recovering some proportion of the sum owed to him for his work for the company. (For example, on 9 July 1845 he had presented a personal account including tradesmen's bills of £3092 3s 4d and £1500 for expenses.)

A result of the South Eastern Railway's partial success was that the Bridge Wardens obtained their Bill for Cubitt's road bridge. Barely ten years later an East Kent Company was after all allowed to build a second bridge, designed by Cubitt's son, Joseph, for a line which ultimately became the London, Chatham & Dover Railway.

On 6 April 1845, Vignoles wrote in his diary 'Rested all day at Manchester after 18 or 19 days knocking about.' He had just completed a round trip which had taken him from London to Leeds, Edinburgh, Dumfries, Liverpool and Dublin, thence via Belfast to Scotland again, and so to Manchester. A memorandum of charges and expenses to be made to twelve different railway companies completed the entry, the total being £580, with the comment 'Profit equal to £30.0.0. per day'. On 7 April he attended a meeting in Manchester of the provisional committee of the proposed North Western Railway, at which he recommended an increase of capital from £1 000 000 to £1 200 000. That night he travelled back by rail to London, in time to give evidence during the next two days at the House of Commons enquiry into the atmospheric railway.

Yet in the midst of all this activity he was considering a step which could have removed him from England for several years. For on 8 April, besides attending the House of Commons committee, he was interviewed by the Chairman and other officials of the East India Company, as candidate for a three-year post as consultant engineer to report on the possibilities of an Indian railway system. Vignoles was attracted by the proposal, though he pointed out that it would involve him giving up 'a large income and prospects of money and honour' in England. Captain Beaufort, Goldsmid and Palmer (Vignoles's solicitor) and other friends advised him to accept the post if it was offered, but only on the most favourable financial terms. At a further meeting at East India House, according to Vignoles's diary, the affair was nearly settled, and he began looking around to find other engineers who might be willing to take on his English work. But when the Company finally offered him the post on 2 May, he declared he would only accept it at a salary of £7000 per annum, with an additional £1000 for an assistant. As the Company was not prepared to go as far as this, the negotiations came to an end.[3]

While it is interesting to speculate on what might have been the outcome of so great a change in the direction of Vignoles's career, the incident clearly shows how far he had recovered confidence in his future prospects, and the confidence of his friends and employers, since leaving the country for Württemberg only two years before.

The peak of the 'Railway Mania' had come at the end of 1845. The plans of the North Kent line, deposited by Vignoles at the end of November, were among those of at least a dozen projected railways on which he and his staff had been working in his Trafalgar Square office. His diary for 29 November records 'N.B. We did not get to bed for several nights of this week, working incessantly.' This was the occasion quoted by several railway historians, when Hutton Vignoles fell asleep at his desk, and as no one could wake him his Wellington boots were cut off and he was rolled into bed in his clothes. Olinthus Vignoles tells the story as a quotation from the diary, though in fact there is no reference to it there. No doubt his father or his brother told him of the incident.

On 1 December Vignoles admits for once to being 'Quite exhausted with the great exertions made during the last few days.' He allowed himself two days' rest at Brighton before travelling to Rochester for a meeting of the Bridge Wardens, and two days later he was in Lancashire.

Here he was in familiar country, involved with a group of three lines in the northern hills and dales not far from the terminus of his North Union Railway. These were designed to bring the benefits of railway transport to the Lancashire mills, while continuing the urge northward exemplified by the Preston to Lancaster line, of which Joseph Locke was the engineer. The first, the Blackburn, Darwen & Bolton Railway, like the Sheffield & Manchester, crossed a ridge with a steep climb at either end (the maximum gradient being 1 in 70) and a tunnel through the summit. The Blackburn, Clitheroe & North-Western Junction Railway was to all intents and purposes a continuation of the first, following the south-eastern flank of the Ribble valley north-eastward from Blackburn, to join the Leeds & Bradford, and Lancashire & Yorkshire lines as they pushed up to Skipton. The third, the North Western Railway, was to run north-westward from Skipton by way of the Wenning and Lune valleys to Lancaster, with a branch from Clapham, through Ingleton, to join the Lancaster–Carlisle railway near the 'Tebay Gap'. (See map, p. 27.)

While Vignoles held the position of Engineer-in-Chief to all three of these lines, and kept as close an eye on their planning as his wide range of commitments would allow, the brunt of the work fell on John Watson, a former assistant of his, with whom he enjoyed an effective working partnership for some years. Watson had made the first survey of the Blackburn, Darwen & Bolton Railway before Vignoles was called in as a consultant, when, in view of the steep gradients, his first recommendation was that the line should be worked on the atmospheric principle. Vignoles was subsequently appointed Engineer-in-Chief, while Watson had the title of Acting Engineer. The arrangement worked well on the whole. It did however give rise to some misunderstandings as regards responsibility, and led to a serious dispute between Vignoles and John Lister, resident engineer to the Blackburn railway. In November 1845 Vignoles severely criticised him for having communicated directly with the Company Secretary, instead of through the channel of the Acting or Chief Engineer. Lister maintained that as he had been appointed by the

Company he must be directly responsible to them. His assertion had some basis, in the uncertain definition of the chain of responsibility between a Chief Engineer who was often inaccessible for long periods, an Acting Engineer, and the resident working continuously on the ground. The Board members were inclined to sympathise with Lister, and did not make things easier by inviting him to submit designs for bridges and a station direct to them. Such action by the Directors naturally infuriated their Engineer-in-Chief, who called for Lister's immediate discharge, and insisted that all engineering plans and details should be submitted for consideration by him before being seen by the Board.

Seeing that they had to choose between Vignoles and Lister, the Directors resolved at a special meeting on 21 March 1846 that ' . . . regretting such differences have arisen, [they] feel they have no alternative but to cancel Mr Lister's appointment as Resident Engineer'. The tone of their resolution was hardly designed to please Vignoles, and relations between him and the Company continued to deteriorate.

Six months later the Company found themselves threatened with legal proceedings, owing to a deviation having arisen in the line of the Whittlestone Head tunnel, beyond the limits allowed by the Parliamentary plans. They put the blame for this on Vignoles's assistant Cooper, whom they asked Vignoles to dismiss, and informed him that they would hold him responsible for any damages they might be charged. As security against any possible claim, they proposed to withhold payment of £2000 owing to Vignoles as fees and expenses. It is not surprising that, although admitting some responsibility for Cooper's error, Vignoles now sent in *his* resignation. Watson remained as acting engineer, and one of Vignoles's former pupils from Ireland, Terence Woulf Flanagan, was to complete the line under his direction. Flanagan was also responsible for building the Blackburn & Clitheroe line, after it had been laid out by Vignoles and Watson. The third of these northern lines, the North Western (known in railway history as the 'Little' North Western, to distinguish it from the L.&N.W.R.), we shall consider in a later chapter.

Meanwhile in Ireland, since the failure of the Irish Railway Commission's recommendations, private promoters were on the move, though the scale of their activity was small compared with what was going on in Britain. Vignoles had his share in this activity. Although the Grand Canal Atmospheric Railway Bill was thrown out in favour of the Dublin & Cashel Railway Company, he was pleased by the success of the Kingstown & Bray line. Further south, he had since 1841 been nursing the idea of an east–west link across southern Ireland, between the ports of Waterford, in the south-east, and Limerick, at the head of the Shannon estuary. The Waterford & Limerick Railway Company was formed early in 1844, and Vignoles was appointed Chief Engineer in October of that year. About the same time he was also engaged as consultant to proposed lines between Cork & Bandon, and Cork & Limerick.

Vignoles laid out the general line the railway was to follow, and assigned the preparation of the Parliamentary plans and sections to a survey team headed by Charles Forth, one of his first assistants, who was now

County Engineer in Limerick. Forth was joined by young Charles Vignoles, who was about to leave his post on the Shannon, and four assistants from England, including Hutton Vignoles, and Thomas Croudace, who was in financial straits owing to the collapse of his coal-shipping business. Vignoles was back in Waterford to inspect the work early in November, and again at the end of the month. By this time the survey was so well advanced that he was able to give Forth instructions for winding up his team. The usual rush to complete the plans for Parliament followed, accentuated by a strike of engravers' employees. Vignoles noted that the crisis was overcome by 'the indefatigable energies of my Staff'. He also records that the final plans were brought over from Ireland by his 17-year-old son Henry, who had been staying with his brothers for a holiday.

The Waterford & Limerick Railway had to compete with some opposition from the Dublin & Cashel Company, which was naturally interested in possible links with southern ports, and claimed the right to establish a line between Tipperary and Limerick. Vignoles lobbied M.P.s vigorously on behalf of the Company, and accompanied a deputation of the Directors to put their case to Sir Robert Peel, the Prime Minister, afterwards circulating an account of the interview to the editors of railway and engineering journals. In the end the Dublin & Cashel Company withdrew its opposition and in July 1845 the Waterford & Limerick Railway Bill was passed in the House of Lords. In August the first general meeting of the Company was held, and on 15 October the first sod was cut by Lord Clare, at Boher, near Limerick. The contract for building the line was granted to Vignoles's old associate, William Dargan.

Construction began with the section between Limerick and Tipperary, for which the resident engineer was Richard Osborne, a well-connected Irishman with considerable experience in North America, where he had been Chief Engineer to the Philadelphia & Reading Railroad Company. Osborne has left an interesting account of his work with the W.&L.R. in a so-called diary, in fact à series of recollections written in old age. In comparison with contemporary records he is often extremely inaccurate, particularly as regards dates. But the story he tells of how he came to be appointed to the W.&L.R. is worth quoting, for the fact that it is a rare account of an actual meeting with Vignoles. Osborne writes:

. . . on morning of 22 May 1845 I called on Mr. Vignoles; he received me pleasantly. I made known who I was, where I had been engaged, gave my references, talked of the Osborne ancestors etc. Mr. Vignoles listened and began at once to question me about bridge work, and asked if I could build a 400 ft. span. I said, Yes, the very last bridge I had arranged before leaving America was a span of this length for . . . the Philadelphia, Wilmington & Baltimore Railroad.

Mr. Vignoles seemed pleased, and at once said that he would like me to make a sketch of such a bridge as I would propose. 400 ft span with the rails 90 ft. over water level. I immediately left to carry out his wishes. I soon completed a neat general plan, and took it to Mr. Vignoles on 27 May 1845.

He had a large force of assistants and draughtsmen, getting up the plans and documents for his Parliamentary examination for the charter of the Leeds & Manchester Railway, and was hard pushed to get through in time.

He looked at the bridge drawing I had brought him, made no comments or inquiries, but remarked that his son had a large force in Ireland, getting up the Parliamentary plans for the Waterford & Limerick Railway, and after asking if I had ever been in Ireland, and being told my family estates were in the Counties of Tipperary, Kilkenny, Waterford & Wexford . . . he said 'Well, I will get you to go and relieve my son in Ireland, and then I will appoint you my resident chief to construct the road.'

Almost too astounded to speak, I managed to express my thanks, and my appreciation of his confidence in me, and asked to be permitted to aid him in perfecting his Leeds & Manchester plans, which I would gladly do if he would introduce me to his Principal above stairs, to which Mr. Vignoles gladly assented, and I was soon to work there. . .

Osborne goes on to describe his journey to Ireland, in the course of which he was gratified to make the acquaintance of the Duke of Leinster. On arrival at Limerick 'I lost no time in delivering my letters to young Charles Vignoles (who was glad to be relieved from duty there) and was soon at work in my new station, and pushing our Parliamentary work.' According to Osborne the work was completed by the middle of August 1845, when he returned to London to submit the plans to Vignoles, after which the case proceeded to Parliament.

Leaving aside the obvious inaccuracies in Osborne's dates, his account of his interview with Vignoles has the ring of truth. So does his description of the office at 4 Trafalgar Square. (He can perhaps be excused for inventing the name of a railway Vignoles never planned.) In other respects his story is difficult to reconcile with contemporary records and the day-to-day account in Vignoles's diary. Vignoles could well have been attracted by Osborne's account of his career, in some respects similar to his own, and the quick decision to offer him the post of resident engineer is quite in character. But the first reference in the diary to Osborne's appointment as resident engineer is dated 11 July 1845. It is interesting that Vignoles mentions on the same day a long spell of work on the preparation of a 'memorial' for the Board of Trade on a 400-ft timber bridge over the River Suir at Granagh near Waterford. This was undoubtedly the bridge to which Osborne refers. However, if this was the occasion described by Osborne, we have to discount his claim to have been appointed to complete the plans for Parliament; if we are to accept that he completed the Parliamentary plans, then his meeting with Vignoles must have taken place in the autumn of 1844. The most likely possibility is that he took over from young Charles Vignoles late in 1844 to work with or under Forth, rather than himself taking charge of the survey, as his account suggests; and that he was appointed resident engineer in July 1845, a few weeks after Forth's sudden death.[4]

The work of building the Limerick to Tipperary section proceeded well under Osborne's direction, and in December 1845 the Directors showed their confidence in him by voting him a salary of £800 per annum plus expenses, as General Superintendent of the line. At the same time they offered Vignoles £1000 per annum, conditional on his visiting the line at least eight times a year. Vignoles, who was much preoccupied with the North Kent line and his other work in England, would have preferred to make only six visits for £1000 or eight for £1200. But the Company was now running short of money, and in February the Directors decided to apply for Government aid. At the same time there were difficulties with Dargan about his contract, which Vignoles had tried to smooth over, and disagreements between Vignoles and Osborne. Vignoles had given orders to change some of the gradients laid down in Dargan's original contract. Osborne objected to this, and the Directors supported him, giving him full authority to carry out the work as planned. Osborne's journal describes a meeting with Vignoles (which he dated as occurring on *1 October 1847*!) which ended in the latter presenting the Board with an ultimatum, saying that Osborne must leave or he must. According to Osborne the Board refused to dismiss him, and Vignoles was obliged to resign.

Vignoles's version of the affair is rather different. He records that he spent several days in April (1846) going over the works with Dargan and Osborne, and that at the Directors' meeting on the 16th he protested angrily at their decision to cancel his orders changing the gradients, and to give Osborne full control. The next day, however, he agreed to the changes, and declared he would work for the Company as long as the 'cutting-out' of the line continued, in view of the critical state of their affairs.

Vignoles's account is largely borne out by the Company records. In the absence of government aid the work was stopped beyond Tipperary. On 1 June the Board accepted Vignoles's resignation without acrimony. The work from Limerick to Tipperary was ultimately completed under Osborne's direction, though at first only a single line was laid. The irony of the situation lay in the fact that a vigorous programme of railway building, backed by English capital, could have done far more at this time to alleviate the miseries of the Irish potato famine and of Irish unemployment, than the half-hearted public works programme which the Government organised.

The story of the relations between the W.&L.R. Company and their Chief and Resident Engineers follows a pattern not unknown in engineering history. As we have seen it was to be repeated almost identically a few months later in the case of the Blackburn, Darwen & Bolton Railway. There is no doubt that the W.&L.R. Directors believed that Vignoles was not giving enough attention to directing the building of their line. But in fairness to Vignoles one must point out that he was not the only engineer to find himself in this predicament. The fact was that the enormous expansion of railway promotion had outstripped the resources of the profession. There were simply not enough engineers to go round.

Vignoles stretched himself to the limit. It was typical of him that when he was in Ireland for the first general meeting of the W.&L.R. Company, he had with him the plans of a 'harbour of refuge' at Dover, on which his assistants at Tipperary worked for a day and a half, while he himself was completing a report to be submitted to the Admiralty Hydrographer, Captain Beaufort, a fortnight later. He rarely refused the opportunity of new work, partly because of his boundless energy, but also because his experience with the Sheffield & Manchester Railway had left him a legacy of anxiety about money which was to dog him for many years. The upkeep of a large staff and offices was a constant drain on his resources, which could not always be met by the somewhat irregular payments he received from his employers. Meanwhile he was by no means free from private and domestic worries, many of them financial. In May 1845 his son-in-law Croudace went bankrupt for a second time, thereby involving Vignoles in a loss of £1500. He also proved to be unfaithful to his wife. Vignoles was obliged to provide for Camilla who, at the age of 27, with her two youngest children, took refuge in France with friends of her schooldays, where she planned to eke out the allowance her father made her by giving lessons in painting; at the same time he provided for the education of the two elder Croudace boys, Charles and Walter.

The health and temperament of his eldest son gave Vignoles further cause for anxiety. Damp winters on the Shannon had given him chronic rheumatism, and after the completion of the W.&L.R. plans Vignoles had suggested that he should return to England. Unknown to his father and 'not at all in a manner to give me satisfaction' he had married while in Ireland. For a few weeks he and his wife stayed with the family in Manchester, where the even keel of life at 24 Mill Street was upset by a violent quarrel between Charles and Hutton, which ended in blows. The event was fully and graphically reported to Vignoles by Mrs Turner, who laid the blame firmly on Charles's wife, while expressing her fears of the consequences of Charles's uncertain temper. Relations between the brothers remained strained; but fortunately in April 1845 Charles was accepted by Robert Stephenson as assistant on the Leeds & Bradford line. Here he seems to have given every satisfaction, until seven months later he had some kind of fit bordering on insanity, an event which so shocked his father that he recorded it in his diary in French, with the conclusion 'je craigns trop qu'il va terminer ses jours comme sa malheureuse Mère!' It was a fear that was to prove true in the end, though at the time Charles made what seemed to be a complete recovery. He continued to work for Stephenson for several months more, which says something for the friendly relations between Vignoles and Stephenson, and the latter's opinion of Charles's ability.

Young Charles was of an inventive turn of mind. Among the patents he filed was one for 'Vignoles's Steam Railway', about which he had an explanatory pamphlet printed in August 1846. This was similar in principle to the atmospheric railway, but was to substitute steam for atmospheric pressure. The pamphlet was illustrated with elaborate sectional drawings, but Charles was not a mechanical engineer, and lacked the experimental facilities to overcome the practical difficulties with which his invention bristles. There is no record of any attempt to develop the idea.

Although Vignoles sometimes grudged the money Charles spent on patents, he was ambitious for him and for his brothers, and looked forward to their following in his own footsteps. He employed them himself, and believed in giving them responsibility at an early age, though he was apt to be disappointed and hurt if they did not come up to the standard he expected of them. Despite the firmness with which they had been brought up as children, he enjoyed their affection as well as their respect. Unfortunately, however, though Charles and his father had travelled and worked together a good deal in early days, they had seen little of each other since Charles's appointment in Ireland. This and the separation from the rest of the family may have given Charles a sense of grievance, which the family's general disapproval of his wife only intensified, and this may have had an effect on his mental state.

In September 1846, Charles ended his work with Stephenson and joined John Watson on the Blackburn, Darwen & Bolton line; but after only two months the appointment came to an abrupt and unhappy end. Once more he was the victim of his unfortunate temperament; he quarrelled violently with his employer and was dismissed. Vignoles was out of England at the time. A ten-week gap in the diary, and two letters written from Marseilles suggest that the strain of these years of struggle had at last begun to tell, and that he had been forced to take the first real holiday of his life. If so, the news of Charles's dismissal must have been a bitter interruption. In a hastily scrawled letter to Watson dated 3 November from Marseilles, he expressed his regret at Charles's behaviour. He himself was so grievously hurt by it that his nerves were shattered, and he could scarcely hold a pen. His health and spirits were upset by the turn his many affairs had taken, particularly the dispute over the Whittlestone Head Tunnel, and the necessity of going to law to obtain what he was owed by the North Kent and other companies. He begged Watson to keep things quiet for a time, since he felt unable to return to England for some weeks.

Weighed down by sheer fatigue, anxiety and frustration, it seemed that for the moment Vignoles's perennial optimism and ebullience had deserted him. His spirits were at their lowest ebb since the Sheffield & Manchester disaster. Yet within six weeks he was to be on the threshold of an undertaking which would prove to be the greatest achievement of his career.

10 The Kiev Bridge I, 1846–1849

If circumstances had allowed Vignoles successfully to complete the Sheffield & Manchester Railway with its formidable Woodhead Tunnel, this would surely have placed him alongside his three great contemporaries, I. K. Brunel, Robert Stephenson and Joseph Locke. As it was, while each of the 'Great Triumvirate' had a major work to his credit, Vignoles's achievements so far were small in comparison with theirs, in spite of the vast amount of time and energy he expended on the work he undertook; and he had to wait seven years after leaving the S.&M.R. before he had the opportunity of tackling and completing a project on a comparable scale. This was to be not a railway but what was at the time the longest suspension bridge in Europe.

Among Vignoles's many foreign contacts was the British Consul General in Warsaw, Colonel Du Plat, with whom he had some years before discussed the possibilities of railways in Poland and Russia. During the autumn of 1846 Du Plat wrote to Vignoles that the Tsar Nicholas I was requiring an engineer to build a bridge over the River Dnieper, at Kiev. Vignoles must have had this project in mind while out of England, for it is mentioned without preamble, in an entry in his diary on 12 December, the day after his return from France. By this time it appears that he had received particulars of the site from Du Plat, and an assurance of support from Douglas and Alfred Evans, two Birmingham engineers and businessmen who had interests in Warsaw.

The Tsar Nicholas was a military autocrat, who ruled his vast territories like an army, rigorously suppressing any hint of liberalism, and doing his best to keep Russia isolated from the dangerous political influences of Western Europe. Yet despite his opposition to revolutionary tendencies from abroad he was none the less steeped in that international culture of the European Courts, of which the hallmark was the use of French as a common language. He had observed the benefits of European industrial progress, and realised that, owing to her extremely backward state of development, Russia could only enjoy them with the help of foreign engineers and foreign capital. He had in 1840 sent General Tcheffkine, his Director-General of Roads and Bridges, to England, when Vignoles had acted as his guide. Since then, American engineers had begun to build the Moscow–St Petersburg Railway. No doubt, when Vignoles's name was mentioned in connection with the Kiev Bridge, Tcheffkine was able to assure the Tsar that in engaging him he would not be appointing a revolutionary firebrand. Besides, Vignoles was a soldier, and could be expected to understand the importance of siting the bridge where it could be dominated by the guns of the fortress of Kiev.

Once he had made up his mind to apply for the contract, Vignoles's recent depression seems to have vanished, and he moved with his usual swiftness. He was faced with the problem of bridging a river nearly half a mile wide, subject to severe flooding after the melting of the winter snows, with consequent frequent changes in the sandy river bed. For this reason the number of piers and spans had to be reduced to a minimum; an arched bridge of whatever material would be difficult to build and very expensive; he therefore decided on a suspension bridge of four or five spans. He was familiar with the Menai Bridge, and had worked on suspension bridges with Captain Brown and Tierney Clarke in his first years as an engineer. He had long discussions with Clarke, searched the library of the Institution of Civil Engineers and studied all the plans of suspension bridges that he could lay hands on. Having made the important decision to use Captain Brown's system of suspension by chains made up of flat wrought-iron bars bolted together (as used by Telford in the Menai Bridge), he prepared drawings of his proposed plan, and commissioned the artist John Cooke Bourne to make perspective views of the bridge in water-colour. These, together with his plans, he had mounted on silk and bound up in portfolios of red morocco leather. From the first he was determined to travel to Russia and deliver his plans to the Tsar in person.

Preparations for the journey included a confidential interview with Lord Palmerston, the Foreign Secretary, who provided Vignoles with sealed letters of introduction and requested him to keep his eyes open for items of political interest to the Foreign Office. Legal actions pending against the two Blackburn companies and the North Kent Railway Company for balances due to Vignoles were left in the hands of his solicitors; but money for his journey had to be raised by loans on other sums he was owed. The capable John Watson remained in charge of current work, though Vignoles did contrive a quick journey to Settle and back to be present at the cutting of the first sod of the 'Little North Western' and at the usual ceremonial dinner and 'subsequent festive proceedings'.

He found time to report to Beaufort and Burgoyne, and to Admiral Sir Byam Martin, on the use of concrete blocks which he had observed in the harbour works at Marseilles and Algiers during his trip abroad, and to recommend that similar blocks should be used at Dover; and he ate Christmas dinner in his Trafalgar Square chambers with his family and James Pim.

Somehow all this and much else was done, and recorded in an almost illegible hand in the diary; and on 3 January 1847 (just over three weeks after his return to England) Vignoles set out, with his sons Hutton and Henry, on the first lap of the long journey to St Petersburg.

Anyone less used to travel than Vignoles might have been daunted by such an undertaking. He did not even take the shortest route, but travelled up the Rhine to Stuttgart, and thence by way of Munich and Vienna to Leipzig, Cracow and Warsaw. The reason for this zig-zag sweep through southern Europe lay partly in his interview with Lord Palmerston before he left London, for one of the subjects on which the Foreign Secretary required information was the possibility of a direct overland rail-route for mails from India. Vignoles had a long discussion on this with Lord Ponsonby, the British Ambassador in Vienna. His route may also have been dictated by the method of travel and the political situation. He had acquired a comfortable travelling carriage for the journey. Where railways existed the carriage, complete with passengers, was put on a train. Otherwise the travellers posted day and night, changing horses at the end of each ten-mile stage. There had recently been a rising in Austrian Poland, and this may have decided Vignoles to take advantage of the railway from Vienna to Leipzig, and to travel thence by road to Cracow, instead of posting direct through a politically disturbed area of country.

He obviously enjoyed the whole journey. In Stuttgart he was warmly received by old friends, and assured that many people regretted, now their railway was being built, that his plans had not been adopted. He showed his bridge designs to the King, as well as to Lord Ponsonby in Vienna. There he drove with his sons in the Prater, and took a stage-box at the Joseph Stadttheater to hear Jenny Lind sing in *La Fille du Régiment*. Every stage in the journey was carefully recorded in his diary.

After entering Poland, he found the names of the towns and post stations to be 'perfectly unwritable, unreadable, and unpronounceable'. Though fluent in French and German, he spoke neither Russian nor Polish. However, with the aid of a good map and generous pay for the postillions, the party made good progress, though reduced one day to asking the way of a parish priest, with whom they could only communicate in writing – in Latin. On 22 January they reached Warsaw, 19 days out from London.

Here Vignoles made contact with Du Plat and the Evans brothers, and dined with Field Marshal Prince Paskievich, the Russian military governor. The latter was at this time ruthlessly enforcing the policy of extreme 'russification' of Poland ordered by Nicholas I after the uprising of 1830. Vignoles was not unsympathetic towards the plight of the Polish people, but he did not let his feelings interfere in his relationship with his future employers. He described Prince Paskievich's dinner as a 'small agreeable party – quite the reception of one gentleman by another'; and his only criticism of the Prince was that he declined to talk about bridges, especially the new suspension bridge over the Vistula at Warsaw, which Vignoles judged to be a very slovenly affair with poor foundations.

He and his sons had now covered some 1500 miles of their journey. On 27 January they set out on the last 700 miles to St Petersburg, taking Alfred Evans with them. As they travelled north the weather became colder and colder. Because of the snow-bound roads the wheels of the carriage were replaced with sledge-runners, and the team of horses was increased from four to six. By the time they reached their destination on 1 February, Hutton and Henry, obliged to ride behind in the coupé, were in serious danger of frostbite.[1]

Vignoles lost no time in getting in touch with Lord Bloomfield, the British Ambassador, and General Tcheffkine; but it was not until 12 February that he was able to meet Count Kleinmichel, the Minister of Public Works, who was to be his link with the Tsar. Having examined Vignoles's plans, Kleinmichel commissioned designs for four other bridges in important towns; and Vignoles spent several days working on these, until he was received by the Tsar at the Winter Palace.

Nicholas I was an impressive and somewhat awe-inspiring personality, but he knew how to unbend, and he won Vignoles's approval by his courteous affability and his readiness to listen to his explanations. These (delivered of course in French) included the reasons for the superiority of wrought-iron bar suspension chains over wire cables, a new idea to the Russians. The Tsar complimented Vignoles on his plans, and declared himself completely satisfied with the proposals, including those for the extra bridges, which were to be begun when the Kiev Bridge was properly under way.

Having obtained the Tsar's consent, Vignoles despatched Hutton to Moscow and Kiev, to investigate the possibilities of obtaining materials, while he returned to London with Henry, pausing in Warsaw to sign a contract of partnership with Du Plat and the Evans brothers, in which a wealthy Jew called Blomberg was included as contractor. Vignoles was to receive £3000 in consideration of his work as engineer and all expenses; after which the profits were to be divided between the partners in the proportion: 5/16 to the Evans brothers, 5/16 to Blomberg, 2/16 to Du Plat, and 4/16 to Vignoles.[2]

In Berlin Vignoles delivered despatches from Du Plat to Lord Westmorland, the British Ambassador, and on 10 March he was back in London. The journey had taken 336 hours with an actual travelling time of 194 hours. That evening he was at work in his Trafalgar Square chambers.

Apart from much home business that needed his attention, his first concern was to engage William Coulthard and his son to make the working drawings for the bridge. With Coulthard he carefully measured the suspension bridges at Hammersmith and Shoreham, and he took the opportunity of a visit to Scotland to study the bridge at Montrose. On 7 April a welcome letter arrived from Kleinmichel officially notifying him of the Tsar's acceptance of the proposal to build a bridge at Kiev at the price of 1 670 000 silver roubles (£239 167 sterling).

By the end of June the working drawings were finished, and Vignoles was ready to return to Russia. This time he took with him his eldest and youngest sons. In spite of a rather depressing report earlier in the year by an eminent Dublin physician, young Charles was now so much better in spirits that his father thought that the trip would do him good; and he had hopes of employing him as an assistant engineer. They travelled by sea to Hamburg, and thence by rail to the Polish border, where the carriage was once more harnessed for the long cross-country posting to Warsaw and Kiev. This time the travellers suffered from extreme heat. On 12 July, Vignoles had his first sight of the Dnieper at Kiev.

It was a challenging prospect. The city stood on the right, or southern

Water-colour sketch by John Cooke Bourne of the proposed suspension bridge at Kiev, from the portfolio presented by Vignoles to the Tsar in 1847. Compare the elaborate decoration with the more severe style of the finished bridge. Bourne has also added an extra span.

Kiev Bridge 1899. The photograph clearly shows the very substantial girdering which Vignoles used to stiffen the roadway of his bridge (see p. 143). In the illustration the first span is in course of repair.

bank of the river, which at this point flowed nearly east. The Podol, or commercial quarter, was situated on a flat plain at river level, the rest of the city on higher ground to the east, which rose to a steep sandy bluff 400 ft above the stream, crowned with fortifications and numerous public buildings and churches. Prominent among these were the tall towers of the Lavra monastery, capped with their onion-shaped domes. On the left bank lay a flat marshy plain, transformed into a vast lake each spring by the melting snows, across which the only approach was a raised causeway. It was from the foot of the bluffs to the causeway opposite that Vignoles's bridge was to be built, spanning a stream nearly half a mile wide, whose maximum depth varied from 40 ft in autumn to 60 ft during the highest spring floods.

His plan was for a bridge with four principal spans of 440 ft, and two side spans of 225 ft. At the Kiev end a passage of 50 ft spanned by a swing bridge would allow for the passing of the typical high-masted sailing-craft of the Dnieper and for the lofty funnels of the steam-boats which had recently begun to appear on the river. Five main suspension piers were to be erected in the river; one mooring abutment for the suspension chains would be at the outer end of the swing bridge; the other would be built on the northern or left bank. The bridge platform would have a total width of nearly 53 ft, of which 35 ft would form the carriage-way, narrowing to 28 ft where it passed through the portals of the suspension towers, under semi-circular arches 35 ft high from roadway to soffit. The platform was to be suspended by four chains, two on each side of the carriage-way, hanging in the same horizontal plane, and composed of flat wrought-iron bar links 12 ft long, 12 in. broad, and 1in. thick, eight links forming the breadth of each chain. Footpaths on either side would project beyond the chains, and be carried on cantilevers round the outside of the piers, thereby completely separating pedestrians from horses and wheeled traffic. Vignoles proposed to build the piers of brick faced with granite, on concrete foundations for which substantial coffer-dams would have to be constructed. The final architectural design was to be simpler and more severe than the elaborately decorated style of Bourne's sketches, Vignoles declaring that he intended the towers to harmonise with those of the numerous fortresses overlooking the river. In this he was following Telford's example at Conway, though the final result at Kiev was to be less fussy than the pseudo-Gothic style of Telford's suspension towers.

After a couple of days inspecting the site, during which he established the fact that the Kiev iron foundries were not capable even of producing the iron heads and shoes required for the timber piles, Vignoles and his sons took the road again, this time for St Petersburg. Here he was obliged to spend a whole month waiting on the findings of the Bridge Commission, a body of 30 officials deputed to examine and criticise his plans. He had serious doubts of their competence, doubts shared by Tcheffkine and General Gottmann, the Kiev representative of Tcheffkine's Ponts et Chaussées department. St Petersburg was hot, dusty and overcrowded. Charles and Olinthus left for England, and Vignoles was once again consumed with impatience, relieved by a day's sailing off Kronstadt in an

English friend's yacht, and a spectacular military review, conducted by the Tsar in person.

At a meeting on 10 August the Commissioners demanded that the bridge should be made wider, and the masonry facing of the piers heavier. These points Vignoles was prepared to concede, but when they questioned the strength of the suspension chains he was not disposed to agree. However, he spent a long day (rising at 3.30 a.m.) checking all his calculations, and satisfied himself that the proposed weight and sectional area of the chains, and the strength of the vertical rods, would be more than adequate for the load they would have to bear, even with the increases recommended by the Commission. Other meetings and discussions followed. The Commissioners demanded more details, which Vignoles was reluctant to give; in this he had General Gottmann's support, who declared that the plans ought to be accepted without more enquiry, and advised him against signing anything which would admit the right of any commission to interfere with his designs, once the work was begun.

At last, at the beginning of September, he was relieved to hear from Kleinmichel that the Tsar had seen the Commission's report, and had decided to give Vignoles a free hand, conditional on his accepting the alterations to the bridge platform and the piers. (The final dimensions agreed were those given on p. 122 above.) Any outstanding points would be settled directly between him and the Tsar. For this purpose he was conducted by Kleinmichel to the country palace of the Peterhof. Two closely-written quarto pages of his diary describe the visit.

On arriving at the palace they were invited to take coffee with members of the Court while awaiting the Emperor's return from church. Just before twelve a move was made to the palace reception room, where various members of the Tsar's family were assembling.

About ½ p. 12 numerous Servants brought in about 7 or 10 Tables on which elegant Déjeuners à la fourchette were ready set out. At one Table the Imperial family and some of their principal Suite sat down. The rest of the Court stood at the other Tables. Suddenly a Lady entered the room, & came straight up to me & addressed me in English. I soon found that this was the Empress, who spoke of the drawings of *all* the Bridges. . . . After speaking of the visit of her Son the Grand Duke Constantine to England, and paying me one or two compliments the Empress invited me to partake of the Breakfast and approaching the nearest Table called a Servant to help me. The Grand Duke Constantine & His Brother & several of the principal Courtiers also came up to me in succession – and spoke of the Designs for the Bridges, which evidently had made a great impression. About 1, the room was cleared of Breakfast & the Court Circle broke up – the Empress proceeding with some of her Ladies to take a Carriage Airing – and almost immediately afterwards Count Kleinmichel and myself were called into the Emperor's private room, which we entered together. Of course I was in my Military Uniform. On entering the Imperial Closet, the Emperor advanced and shook me by the hand, expressed himself glad to see me, and made me immediately sit down between H.I.M. and Count Kleinmichel. The various points relating to the Kiev Bridge were then discussed, and the Emperor, after having explained why it was necessary to have the Portals 28 ft. wide, at once frankly and nobly decided to leave everything else to my own Judgement and Experience. 'Si vous voulez me répondre sur votre parole d'honneur que le Pont sera stabilement construit, je vous laisse pleine action, et je vous en donne la main.' I answered without a moment's hesitation & looking the Emperor full in the face 'Sur mon honneur et sur ma tête.' The Emperor then held out his hand & took mine & shaking it heartily said '*It is a bargain.*'

True to his promise, the Tsar met Vignoles at Kiev on 22 September for a discussion of final points. Vignoles had been disturbed to learn that the Tsar objected to Jews residing at Kiev, since it was likely that the greater part of the contracting work would be handled by Jews; and he was relieved to hear later that in this instance the objection had been withdrawn. On 25 September the formal contract was signed. Kleinmichel had expressed the opinion that it would be sufficient for General Gottmann to sign on behalf of the Russians, but Vignoles rightly refused to be satisfied with anything less than Kleinmichel's signature. In the end the contract was duly signed by Kleinmichel and Gottmann on behalf of the Tsar. Vignoles left next day for Warsaw, having to hire peasants' horses, since all the post horses had been commandeered for the Tsar and his suite. On 3 October he was back in London.

During the winter of 1847–8 work in England on the bridge materials began in earnest. While stone and brick could be obtained locally, every ounce of iron and every item of machinery, such as pile-drivers and pumping-engines, would have to be shipped out from England. Vignoles placed orders for the suspension chains and rods with Fox, Henderson & Co. of Birmingham, and for the rest of the iron-work with Musgrove & Sons of Bolton. The cargoes were consigned from Liverpool by sea to Odessa, from where they would have to travel 400 miles to Kiev by ox-wagon. The freezing of the Black Sea added to the difficulties of the journey. On 2 December Vignoles was at Liverpool personally seeing to the stowing of the first consignment of machinery in the cargo-ship *Flirt*, and promised the captain a bonus of £20 if he made a quick voyage to Odessa 'without being frightened of a little ice in the Black Sea or in the Port itself'.

In view of the Bridge Commission's criticisms he now enlisted the help of mathematician friends in London to assist him with exhaustive checks of all his calculations. While on a visit to Dublin he also submitted them to Dr Romney Robinson, Professor of Astronomy at Armagh, to whom he had been introduced by Bergin, and to a brilliant young mathematical scholar of Trinity College, Dublin, Edward Whiteford, who was persuaded to join the team of assistants being recruited to work at Kiev. Meanwhile young Charles and Hutton were assisting William Coulthard with his drawings at Preston. Charles made sketches of the boring apparatus at Woodhead for despatch to Douglas Evans in Warsaw, and Hutton was responsible for detailed diagrams of the dimensions of the stones to be used in the abutments, which were sent off with careful instructions to the Russian stonemasons.

By the beginning of February a second shipment of machinery had left Liverpool, and to Vignoles's relief the first payment of over £1000 had been received through the Russian Embassy. He made arrangements for a long absence from England, and for setting up house in Kiev, for which all the necessary glass, china, linen and other items, including such luxuries as water-closets, had to be conveyed from England. To his great regret young Charles was left to work with Coulthard at Preston, as he was not thought to be fit enough to go to Kiev. Olinthus had passed into Trinity College, Dublin, where he was to read for holy orders; Henry had found the classics papers too much for him and, instead of entering the university, he joined Hutton at the head of the staff of assistants to go to Kiev; and as a result of Thomas Croudace's continued neglect of his wife, which threw her once more on her father's resources, Camilla agreed to keep house for the party, taking two of her children, Charles, aged ten, and Camilla, aged four, with her.

Somehow, after various vicissitudes on their respective journeys, the establishment was duly assembled at Kiev. Vignoles's household was completed by a chef, a valet and a lady's maid engaged in Warsaw. The engineering staff, for some of whom additional accommodation had to be found, was made up as follows: Chief Engineer: Charles Vignoles; Assistant Engineers: Hutton and Henry Vignoles, John England and John Coulthard (William Coulthard's son); General Assistant: Edward Whiteford; Engineer in charge of landings at Odessa: William Dacre Wright; Mechanical Engineers: John Coulishaw and William Bell; Blacksmith: George Pemberton; Engineer in charge of coffer-dams: Daniel Frost, assisted by his son, Daniel Frost Jnr.

On Sunday, 27 February, the whole party attended Divine Service in Vignoles's house. Two English ladies resident in Kiev supported Camilla

The bridge works at Kiev. A photograph by Roger Fenton, probably taken in 1852, and the only example of his bridge photographs that the author has been able to trace. The view is from the left bank of the river, when the water was very low. At the left a typical sailing-barge, aground on a sand-bar; the Lavra Monastery on the heights; on the right, the inclined plane for bringing materials down to the river level; the first portal (river pier number 1) and the abutment for the swing-bridge are on the extreme right, partially concealed by scaffolding; the rest of the bridge is out of sight to the right.

at the service. Next day the new arrivals were introduced to the leading residents of the city at a ball given by the civil governor. A formal exchange of calls followed, and soon the whole family and engineering staff were established in Kiev society.

The country was still in the grip of winter. Odessa was ice-bound, and the *Flirt* was held up at Constantinople. But at Kiev the wintry conditions greatly eased the task of setting out the exact line of the bridge, since the surveyors could walk and place their instruments anywhere on the ice, and take soundings through it to establish the depth of the river.

Before a beginning could be made on the bridge foundations, a temporary timber bridge on piles had to be erected. With the coming of the thaw in March, the first pile was driven home by manual labour, no machinery having yet arrived. Dacre Wright, at Odessa, had encountered difficulties only to be overcome by a moderate greasing of palms; however, on 13 April, the first train of 12 ox-wagons arrived, after 12 days' journey from Odessa, carrying 507 pieces of engine-parts. Two more trains, of 22 and 15 wagons, arrived on the 18th and 19th. Once the engines were assembled and mounted on barges, work could begin on driving the piles for the temporary bridge. In all, nine steam-engines were to be employed on the works, two of them static, and developing up to 50 hp, while the smaller engines of 4 to 8 hp could be moved from place to place as required. Some of these engines, which were used for pumping water, grinding cement, hoisting timber, stone and iron, and drawing heavy loads, were made by Gough; while James Nasmyth supplied pile-driving engines of a type Vignoles had seen in operation on the site of the Tyne Bridge being built by Robert Stephenson at Newcastle.

Meanwhile supplies of timber and stone began to arrive. The timber was floated down the river in enormous rafts, the stone carried down in sailing barges from quarries upstream. The principal supply of granite was eventually to come from nearly 100 miles away, and was transported by bullock-cart through rough country, almost devoid of roads. Work on the temporary bridge proceeded slowly. Some of the timber rafts grounded on sandbanks owing to an unusually rapid fall in the river level. The week's holiday taken by the workmen at Easter caused further delay, a foretaste of many similar interruptions arising from the Russian peasants' strict observance of saints' days. Vignoles chafed at the loss of fine weather, and of the opportunities for pressing on with the work while the water was low. He was further dissatisfied by the poor quality of foremen and workmen provided by the contractor Blomberg, and by the inefficiency of his agent Schweitzer. There was a particular shortage of men capable of handling the engines. There was also a lack of sturdy small boats. Two of the young assistants were nearly drowned in the Dnieper by the capsizing of a flimsy craft, and Vignoles gave orders for a stout boat to be built to his design, a work which the 'young gentlemen' were happy to take on in their spare time, or when the workmen were on holiday. The boat was launched with due ceremony on Camilla's birthday, 2 May, and named after her.

Borings made on the Kiev side of the river had encountered sufficient

hard clay to provide firm foundations for the swing bridge and the mooring abutment. But borings in the river bed and on the left bank revealed
a very deep layer of sand, into which the coffer-dam piles would have to
be driven as far as possible, in the hope of reaching harder ground. There
were to be eight coffer-dams for the foundations, numbered, for the purpose of identification, from south to north. Numbers 1 and 2 were for the
abutments at either end of the swing bridge, the outer one of which was
also to serve as anchorage for the suspension chains; numbers 3 to 7 for
the five river piers, number 8 for the suspension chain abutment at the
northern end of the bridge. The suspension piers were similarly numbered, number 1 pier being constructed on number 3 coffer-dam, and so
on. The dams were to be constructed of two concentric rings of piles,
firmly braced together with ties and walings, the space between the rows
being packed with clay and straw puddle.

On his fifty-fifth birthday Vignoles made his usual annual assessment
of his condition. It was a happier and more optimistic one than he had
been able to write for some years. He was in good health and spirits and
'free from all the debts & responsibilities which for the last 8 or 9 yrs.
have preyed so heavily on me'. He had every prospect of independence
for himself in future years and of laying the foundations of honourable
employment for his sons. There was, however, one qualification to his
optimism. 'It is true that all depends on the Life of the Emperor of Russia
& on the full success of the Kieff Bridge – but everything at present seems
favourable & I must pray to God that all may go right.'

But the summer brought further setbacks to test his patience and optimism. Blomberg had promised that the temporary bridge and the
coffer-dams would be completed before the winter, but among other
causes of delay, the long wagon trains of machinery were still arriving
very slowly from Odessa; and in an unusually hot spell in the middle of

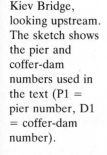

Kiev Bridge,
looking upstream.
The sketch shows
the pier and
coffer-dam
numbers used in
the text (P1 =
pier number, D1
= coffer-dam
number).

June cholera broke out in the city and among the workmen on both banks
of the river. Vignoles commented: 'No wonder! when we consider the extreme state of filth in which the lower classes live – the poverty of their
food and the dampness of their dwellings.'

Fortunately he had moved his family, a few weeks earlier, out of Kiev
to a house he had rented in the country. By 1 July, though the epidemic
seemed to be abating, nearly all the 400 workmen had gone home, and
soldiers had to be brought in to work the pile-drivers. The assistants continued at work. In the ravine just east of the bridge site, Henry Vignoles

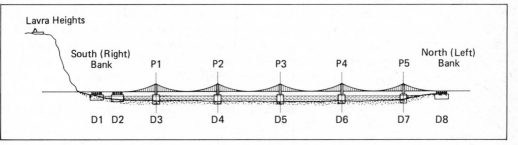

and Whiteford were busy laying out the line of a railway on an inclined plane down from the heights to connect with the railway on the temporary bridge. Vignoles was about to leave to deal with urgent business in England, leaving Hutton in charge of the works. But early on the morning of 2 July Whiteford was attacked by cholera, and in spite of the best medical help Kiev could provide, he died shortly before midday. The little community was deeply shocked and alarmed by this event, especially the two Vignoles brothers and Camilla. Their father wrote: 'The whole family extremely hurt and depressed with deep feelings of anxiety and sorrow. Before going to rest we had Family Prayers in the Church of England Evening Service. I was much struck and affected by the excellent character and disposition of my two sons and all the English Assistants I had with me on this melancholy occasion.'

Mercifully no further cases of the disease occurred among the members of the staff, and shortly after Whiteford's death the situation was sufficiently improved for Vignoles to leave. On his way to Hamburg he shared a first-class carriage for a short time with the Grand Duke of Mecklenburg, 'an affable young man with whom I had a good deal of interesting conversation'.

While in England Vignoles commissioned the engineer Jabez James to make a detailed scale model of the bridge. He also checked Coulthard's final drawings, and with him inspected the suspension chains being made at Fox Henderson's works. The inspection was the occasion of a bitter personal blow to Vignoles. The work was being supervised by young Charles, who flew into a rage when his father made some criticisms of the finish of the links. This was the prelude to an unhappy decline in Charles's mental condition. At the beginning of September he was in so bad a state that the doctors summoned by Vignoles insisted on his being placed in an asylum. From then on, although enjoying occasional periods of lucidity, he was to pass the rest of his life under some form of medical supervision and restraint.[3]

Vignoles was back in Kiev in October. Much progress had been made. The temporary bridge was in use, the coffer-dams completed; and on the heights on the Kiev side several acres of ground had been prepared to receive granite, bricks, timber, cement and other materials which were to be sent down the self-acting inclined plane to the works on the river; on the further bank a whole colony of workmen had been established in huts and cottages specially raised above flood-level. In order to have a continuous record of the work, Vignoles had engaged John Cooke Bourne to make drawings and paintings, and to take photographs on the newly-invented calotype system. Meanwhile he had taken another house in Kiev, and spent a good deal of time with Camilla deciding on decorations and the upholstering of furniture he had brought out from England.

Winter that year began early, and with it came the first test of the contractors' work. In the middle of November large ice-floes filled the river, bearing down on the coffer-dams, and on the piers of the temporary bridge. On the 18th the river was nearly blocked; two heavy barges, driven by a strong wind, instead of passing through the central opening, where a section of the bridge floating on pontoons could be pulled aside, became entangled with the piers and ice, and brought down three or four bays of the bridge. On the 22nd the whole river was full of ice pressing against the bridge, and Henry had a narrow escape. Near the centre of the river a heavy ice-floe striking against the bridge caused him to lose his balance on a loose plank and fall 20 feet into the icy river. 'By the great mercy of God he was able to swim to one of the fast piles and after a few moments' mortal struggle got safe onto the Coffer Dam – with but little hurt.' To close the central passage where floes were still on the move Vignoles had straw and planks laid on the ice and watered; next morning it was once more possible to walk right across the river, and for the time being the threat to the works was over. The real test was to come with the 'débâcle' or melting of the snows the following spring.

It was a heavy winter, in the midst of which Vignoles made a difficult journey to London and back. He was concerned about financial problems which were now beginning to threaten the whole project. Kleinmichel's transmission of payments to the Russian Embassy, for materials purchased in England, were irregular and fell short of the claims Vignoles had submitted. In Warsaw Evans and Blomberg were equally short. Moreover the partners were in disagreement over the amount of expenses to be allowed to Vignoles towards the upkeep of his considerable establishment. He now had to raise a personal loan of £15 000 from his solicitor, Robert Palmer, to maintain the flow of materials and supplies from England.[4] While in England he inspected James's model, arranged about schooling for his two little Croudace grandsons, and had Bourne's calotypes mounted in a portfolio for the Tsar; and on each of his journeys he carried Foreign Office despatches from and to Berlin and Warsaw.

On his return to Kiev on 14 February 1849, he was disappointed to find the work not as far advanced as he expected it to be, and noted in his diary a 'general relaxation of exertions . . . and generally a want of order'. The engines were continually breaking down, a fact which Vignoles attributed to neglect by the contractor, who remained reluctant to use anything but manual labour. The unfortunate Hutton was also a target for his father's disapproval, first for a lack of vigour in his direction of the works, and secondly for his neglect of the accounts of the household over a long period, which came to light when Vignoles was anxious to settle the question of his expenses with his partners. In an entry in his diary on 4 March he blamed himself for giving his son too much responsibility, and he decided that he must engage another resident engineer to relieve Hutton of some of the burden.

Heavy snowfalls in March confirmed expectations that the floods of 1849 would be severe. With the break-up of the ice a tremendous weight of ice-floes piled up against the temporary bridge, finally breaking it in two places over a length of 350 feet. Once the ice had gone, a few fine days followed before the river began to rise. This gave an opportunity to repair the damage. Repair work was well advanced when, at the beginning of May, with the river rising rapidly, the first barges and rafts of timber came downstream, jostling each other for the central opening, and driving down upon the bridge, the coffer-dams and the barges on which the pile-drivers were at work. Vignoles protested bitterly to General Gottman, who issued orders for the control of river traffic which were totally disregarded. Day after day the river continued to rise. On 7 May Vignoles was asked by the Civil Governor to help him to get five regiments of cavalry across the river, by the temporary bridge; but that same day a dozen barges and rafts came hurtling downstream, crashed into the bridge and eventually burst through, carrying away 400 feet of the structure between number 3 and number 4 coffer-dams, while one of the enormous timber rafts was left spread-eagled on the ice-breakers of dams 2 and 3.

Two days later the very existence of the bridge-works seemed to be threatened. On the morning of 9 May, when the tops of the dams were already covered by the flood to a depth of more than five feet, it was observed that dams 2 and 3 had risen two feet above the surface, having been lifted upwards in the sand by the scouring action of the flood. About noon a part of number 4 dam 'shot up like a *whale* about 25 to 30 feet vertically out of the water', and was carried away downstream and grounded on a sandbank, to the great astonishment of General Gottmann, who had just come down to the river to see the damage. Vignoles acted quickly. He assembled all available workmen to fill barges with stones, which were sunk on top of the dams, either to hold them in position or to force them down into their moorings. This saved the rest of the dams, though it could not prevent the remains of number 4 dam breaking away in two pieces and joining the first piece downstream.

There followed two more days of acute anxiety. Then the waters stopped rising, having reached a level 19 feet above that of the previous summer. Vignoles sat down 'with a heavy heart' to write a detailed account of the disaster to Du Plat. Soundings taken round dam number 5 showed that the bed of the river had been scoured out to a depth of 10 to 15 feet. 'A fearful effect of Nature's powers in her wrath' wrote Vignoles. None of the piles had broken, and except in the one case the dams had held well together. But examination of the displaced number 4 dam showed that some of the piles, which had large knots in the wood, had been crushed in driving, and for this reason had not held so well in the ground. Vignoles judged that the undisciplined handling of the barges and rafts was a main contributory factor in the disaster; and although General Gottmann 'now that the mischief is done, croaks about the impossibility of constructing coffer-dams in the Dnieper', he was already thinking how the damage could be made good, and new and stronger foundations laid. This he set out in his letter, under four headings: (1) to add a third ring of piles to each dam, thus providing two clay puddle walls, the clay being mixed with brick and stone chippings, as well as straw; (2) to use piles at least 56 feet long, carefully chosen to be free of knots; (3) to pour concrete both inside the dams and into the scoured channels outside, having first paved the bottom inside the dams with clay; (4) to weight down the dams as heavily as possible with stone. The letter ended with scathing comments on Schweitzer's lack of foresight; for there had not been a single spare anchor or warp in his store, there was no boat (except the *Camilla*) at the works fit to cross the river, and he had not a man who knew how to handle a barge in rough weather.

In spite of his confidence in ultimate success, Vignoles realised that the completion date for the bridge must now be put back for at least a year, a fact he reported to the Emperor in Warsaw at the end of May. At the same time he presented plans of the bridge at Bobruisk, on which he had been working for some time, with perspective sketches by Bourne, only to be told there was no money at present to pay for it.

From Warsaw he returned to England, much exercised in his mind on the problem of the Dnieper coffer-dams.

11 The Kiev Bridge II, 1849–1853

During the summer of 1849 Olinthus Vignoles spent the long vacation visiting his sister and brothers in Kiev. He travelled alone overland, and at Berlin he was asked to take the place of a Queen's Messenger, who was ill with cholera, and entrusted not only with the British despatches for Warsaw, but also with those from the Russian Embassy to Prince Paskievich; a responsible task for 'a young Collegian of twenty summers, who knew only French, with scarcely any German and not a word of Russian or Polish'.[1] At Kiev he tried somewhat unsuccessfully to teach the family to sing partsongs, and when his brothers could spare time from the work of repairing the coffer-dams he would join them and several of the young engineers for an evening on the river, in an eight-oared outrigger which had recently been sent out from England.

On his way home Olinthus met his father in Warsaw and was surprised to learn that he had acquired a stepmother. Without giving any notice of his intention, and without mentioning it even in his diary, Vignoles had married Miss Elizabeth Hodge on 16 June at his parish church of St Martin-in-the-Fields. He does not seem to have broken the news to his family until his return to Kiev in October. Mrs Vignoles did not accompany him to Kiev on this occasion; she had an ageing mother who died shortly after her marriage. But from then on she was a faithful companion on her husband's numerous journeys. Very little else is known of her.

While in England Vignoles consulted many engineering colleagues, among them Robert Stephenson and I. K. Brunel, on the problem of strengthening the coffer-dams. He also wrote to his friend Heinrich Hubbé in Hamburg, who had wide experience of dockyard and river engineering, inviting him to come over to England to see him. On hearing Vignoles's story, Hubbé suggested that he might try the effect of using fascine mattresses, to counteract the scour of the river at the foot of the coffer-dams. This was a Dutch contrivance for the support of piers and embankments, which Hubbé had used successfully on the Elbe. The mattresses resembled large rectangular honeycombs, made of strongly interwoven wattle of willow, ash or birch, which were filled with stone and sunk where underwater support for a structure was required. Here was a possible answer to Vignoles's problem. He travelled to Holland with Hubbé, saw some fascine mattresses in position, and engaged two Dutch engineers to go to Kiev to advise on the best method of using them there. He also engaged a new assistant resident engineer, Abraham Craven, who travelled back with him to Russia at the end of August.

The temporary bridge and the wrecked coffer-dam had been rebuilt and all the dams strengthened according to Vignoles's instructions. But deliveries from Odessa were still slow in arriving, and he had to complain once again of Schweitzer's lack of drive and the constant interruptions of Jewish holidays and Orthodox Church saints' days. However on 14 September he was pleased to note that the concrete foundations in coffer-dam number 8, i.e. the left bank mooring abutment, were complete, and bricklaying had begun in it; while on the 19th the excavations of the swing-bridge abutment were ready for concreting.

Fascine mattresses were now prepared, under the direction of the Dutch engineers, to be laid on the river-bed on either side of each dam. The base of each mattress, 20 ft by 40 ft, was constructed of closely woven and intertwined ropes of brushwood; it was then divided into 4-ft squares by strong stakes driven vertically into the rope-work; after which the cells of the mattress were made by the interweaving of branches from stake to stake in the manner of a wattled fence. The mattresses were floated into position beside the dams and carefully anchored. The compartments were then filled with stone from barges and the mattresses sank to the bottom, their position being carefully checked by divers. Once in position and filled with stone, each mattress weighed about 100 tons.

By 20 October all the mattresses were in place, and the Dutch engineers were paid off. Although the Russians were sceptical about this experiment, Vignoles was confident that it would effectively protect the foot of the dams against the scouring action of the stream. His confidence was to be vindicated with the coming of the floods in the following year. The outer ends of the mattresses gradually sank as the flood swept past them, while the inner sides remained firmly immovable against the foot of the dams, as a permanent bulwark against future spring floods (see diagram, p. 134).

With the coffer-dams safely secured, Vignoles urged his staff to press ahead with the concreting of the foundations before the winter frosts set in. The cement for the concrete, which he described as 'an artificial *poz-zolana*', was made from a peculiar clay found in the Kiev hills, and manufactured according to the principles of a French engineer, Vicat.[2] Large buildings on the heights on the right bank of the river housed eight large roasting ovens and numerous grinding mills, capable of producing 500 cu. ft of cement every day. A leak in number 6 dam caused trouble for a time, but by 28 October it was mastered sufficiently for the concreting to be completed. Pumping was still handicapped by the shortage of engines in operation, since some of their parts had not yet arrived from Odessa. (One boiler which fell off its wagon into a deep pond on the way was left there by the wagoners, and had to be recovered months later by engineers and workmen from Kiev.) Consequently it was not possible to finish all the concreting before the winter frosts set in.

On 3 November 1849 the weather finally broke, and Vignoles once more left Kiev for the winter, after giving careful instructions to the staff, particularly as regards the loading of the coffer-dams before the spring floods. He travelled first to St Petersburg, where he supervised the assembly of the model of the bridge he was presenting to the Tsar, which had been brought out from England by two of James's assistants. Carried out in the greatest detail on a scale of 1 to 96, the 26-ft-long model was almost as great a source of pride to Vignoles as the bridge itself. He personally hunted the St Petersburg shops for figures of people and animals of proportionate scale, explained every detail to the Emperor and worked the swing-bridge for his benefit. The Tsar, who had agreed that the model should be on show in the Winter Palace, was delighted, and deigned to speak a few words in English to the mechanics.

Vignoles was informed by Count Kleinmichel that the Tsar had accepted his designs for the Bobrovisk bridge, and for the other two bridges at Kowno and Dunaberg; but he wished him to submit a combined estimate of cost for the three, payable not in sterling but in Russian bonds. This proposal gave rise to a furious argument between Kleinmichel and Vignoles, who foresaw that it would entail a considerable loss to him, owing to variations in the rate of exchange between St Petersburg and London. Eventually Kleinmichel agreed that the money should be paid in sterling, though Vignoles was obliged to sign an agreement that the final amount should not exceed the sterling equivalent of 2 350 000 silver roubles. Kleinmichel then promised that the arrears of the bridge account due to London and Warsaw would be made up, and the visit was crowned by the Tsar's award to Vignoles of the Russian Order of St Anne.

Diagram drawn by Henry Vignoles, showing the system of cofferdams and fascine mattresses, Kiev Bridge works, 1849.

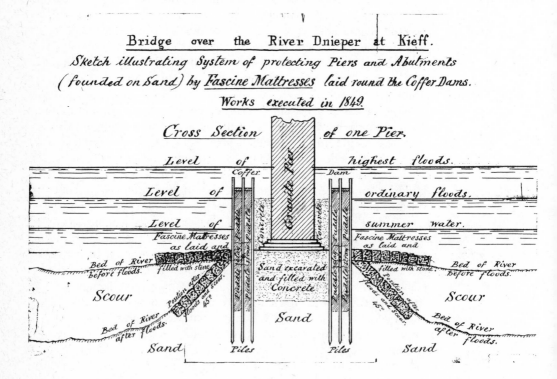

Bridge over the River Dnieper at Kieff.

Sketch illustrating System of protecting Piers and Abutments (founded on Sand) by Fascine Mattresses laid round the Coffer Dams.

Works executed in 1849.

Cross Section of one Pier.

Table 3. *Estimates for the Kiev Bridge works, taken from two*
pages of Vignoles's diary for March 1849

Summary of the Revised Estimates of the Kieff Bridge

16/28 March 1849

A. Expenditure in Russia   Ro.   Sr.   Silver Roubles

1. Coffer Dams & Timber etc. in Foundations.

	Expended	109,489.42	139,489.44
	Remaining	30,000.00	

2. Concrete			63,000.00
3. Brickwork			281,875.00
4. Granite			301,244.24
5. Carriage of Iron & Stores from Odessa			76,500.00
6. Platform of the Bridge, as per Special Contract			60,000.00
7. Erection of all the Iron work of the Bridge			24,000.00
8. Painting, Tarring, Roofing Piers, etc. etc.			30,000.00
9. Incidentals & Extra Day Work		1,016,108	40,000.00
10. Administration, Travelling Expenses & Sundries			85,368.00
11. Professional Advice as per Agreement			18,000.00

Total probable Sum

Ro.   Sr.   1,189,476.66

Contingencies 81,263.00

1,271,739.66

Contingencies.	R. S.		½ Bonus to Blomberg	58,000.00
1. Coffer Dams	3,000.00			
2. Concrete	6,300.00		Ro. Sr.	1,329,739.66
3. Brickwork	28,188.00		Balance of Bonus	53,000.00
4. Granite	30,125.00		Total in Russia R. S.	1,382,739.00
5. Carriage of Iron	7,650.00		Total in England	671,221.24
7. Platform	6,000.00			2,053,950.90

B. Expenditure in England.

£

Less probable
Saving on

1. Iron for the Bridge . . . . .
2. Engines, Tools & Machinery
3. Administration & Sundries
4. All other Materials . . . . .

53,950.00

Contingencies &
Deductions

To cover all Balances to end
Total in England   £100,000

Silver Roubles   2,000,000.00
or say   £300,000 Sterl.
at 3 Shillings per Rouble Sr.

Contractor's Bills   1,074,108 (including ½ Bonus)
Money on A/C

	437,606 ⎫	974,108	Contractor in advance 100,000	See
Further	536,502 ⎭	100,000	Balance of Bonus . 53,000	Contra
Dr.			Ro. Sr. 153,000	

Table 3 (continued).

Abstract of Probable Results of Kieff Bridge Affairs.
16/28 March 1849

C. Receipts (Certain)

	£	Silver Roubles
Cash in London	(116,277.11s.2d)	780,481.75
Probable further sum	( )	122,700.00
Balance in 1852	( )	300,000.00

Total Receivable in London Ro. Sr.		1,203,181.75
Cash at Kieff to January 1849	437,606.00	
Further Payments	709,212.25	
Total Receivable in Russia Ro. Sr.		1,146,818.25
Total, according to Kleinmichel's views Ro. Sr.		2,350,000.00

D. Further Receipts (Probable) Silver Roubles

Re-Sales of Engines, Tools & Mach.	50,000	— 150,000.00
Claim on Russian Government	100,000	

Total Receipts Ro. Sr.	2,500,000.00
Total Expenditure, as pr Contra A & B	2,000,000.00
Balance to Profit Ro. Sr.	500,000.00

Mr. Vignoles' private View of the State of Finances to Complete:–
The Bridge will be finished and all Scores cleared off
for the Sum pr Contra exclusive of Contingencies R. S. ... 1,189,476.66
½ Bonus to Contr. ... 58,000.00

To meet which we have
1. Money already recd. and handed to Contractr.
437,606.00
2. Further Payments in Kieff 709,212.25
1,146,818.25 ... 1,146,818.25
Difference R. S. ... 100,658.41
Add Balance of Contr.'s Bonus ... 53,000.00
Say amount due to Contr. when Bridge is
complete & given up ... 153,658.41
To meet this there will be the advance he undertook to
make in the first instance ... 120,000.00
Leaving only the small sum, which the value of the
Machinery & Iron in possession will amply
meet R.S. ... 33,658.41

No money need be remitted from England at present.

All Vignoles's papers and letters in Russia bear two dates – Old and New Style. The small errors in addition are Vignoles's own, and suggest he was working under the usual pressure when drawing up the table.

But he was sceptical about any improvement in the financial situation. In Warsaw he found Blomberg refusing to engage workmen for the next season unless he was assured of substantial reserves of cash, including a large commission for himself, according to a custom of which Vignoles was becoming painfully aware. While waiting for Kleinmichel's payments, his only resource was personally to raise yet more money at home. This he managed to achieve with the help of his solicitors. At the same time he realised the vital necessity of cutting down his personal expenditure; and with this in mind let his house and chambers in London and rented a furnished cottage for himself and his wife in Tulse Hill at the modest sum of £60 for two years. He also determined to make drastic economies on his return to Kiev.

The spring of 1850 was frustrating. It was now two years since the first pile had been driven, and still there was little of the bridge structure to be seen except the tops of the coffer-dams. The concreting of the foundations proceeded very slowly, owing to the inexperience of the workmen, and was frequently held up by leaks in the dams, which seemed inevitably to be accompanied by breakdowns of the pumping-engines. Vignoles was annoyed to find the pumps in number 2 dam blocked with concrete sucked into them when the frosts stopped work the previous season. He put the blame for this on Hutton, who was already in trouble for having made an error in laying out the foundations of the swing-bridge. On the other hand, he was delighted to find how well the coffer-dams had withstood the spring floods, thanks to the laying of the fascine mattresses and the other precautions that had been taken.

Troubles now arose among the resident staff. The new resident engineer, Abraham Craven, had died during the winter, and to replace him and to ease Hutton's responsibility further, Vignoles had appointed two more resident engineers, a Russian, Captain Kirchenpauer, and Charles Gainsford, an English engineer recommended by Brunel. The latter proved to be less than competent, and considerable friction arose among the staff, tempers being exacerbated by a spell of tropical weather in July and August, when the temperature rose to 100 degrees Fahrenheit. Cholera broke out again among the workmen, resulting in further delays in the work; and most of the English were seriously affected by the heat, and were laid up at one time or another. At the end of August Vignoles had to inform his family of his decision to reduce his household to a minimum, which was to involve sending Camilla and her children home. The situation was hardly improved by Hutton choosing this moment to announce his engagement to a young Frenchwoman called Amélie Le Lorrain. Vignoles refused to give his consent to the marriage before Hutton had finished his contract on the bridge, and only agreed with great reluctance to call on Mme Le Lorrain, about whose antecedents he made careful enquiries.

A visit by the Tsar to the works, at the beginning of October, did something to restore the general morale. All hands had been mobilised a few days before to deal with serious leaks, of sand as well as water, into dam number 3, Hutton and the other assistants and foremen working day and night to overcome the 'inactivity & unskilfulness of the men'. The condition of dams 4 and 5 was also causing some anxiety, and to add to the difficulties a gale sprang up and carried away part of the floating section of the temporary bridge. However, the leak in number 3 dam was brought under control and the bridge repaired in time for the Tsar's visit. The latter declared himself very satisfied with all he saw, though the wind was too strong for him to venture onto the temporary bridge. At a *soirée* that evening Vignoles had the opportunity of discussing with the Tsar plans he had drawn up for improving the bed of the Dnieper by diverting the River Tscheteroi, which entered the main stream some miles above Kiev. He also ventured to explain the difficulties he was having with Count Kleinmichel over payments. The Tsar listened sympathetically and promised that the matter would be put right.

October also brought news of Blomberg's death in Warsaw, which raised further complications concerning the supply of money to Schweitzer and his workmen; none the less, by early November the stonemasons and bricklayers were still making good progress: the granite work in dam number 2 was completed, the brickwork in number 6 up to the bridge platform, and preparations were being made for turning the arch under the roadway in the piers of both 6 and 7. But dam number 3 was still giving trouble; some of the piles had given way and sand and water had flooded in above the concrete already laid at its base. The problem was not merely to pump out the water, but to get rid of the sand which had accumulated inside the dam, before putting in the additional depth of concrete. Vignoles wrote on 24 October: 'We conceive that this may be done in Compartments, separating them by Timber Stands, and using hand dredges and rakes to get out the Sand. When the Concrete is in, to lay a strong outer Course of Granite stones & fill in with Brickwork or Cement.' This method proved only partially successful, and on 1 November Vignoles decided that no more could be done to dam number 3 before the following spring. Instead, he directed that every effort should be made to complete the concreting of dams 4 and 5, and to protect all three dams with an extra row of piles before the winter. He also gave orders that as soon as the brickwork of the abutments was complete the wrought-iron swing-bridge should be assembled and erected, in spite of the mysterious disappearance of a consignment of bolts which had left Odessa a year previously!

He returned to London for Christmas, the first part of his journey being more than usually difficult owing to the appalling state of the roads after heavy rain. At one village the carriage stuck fast in the mud, whereupon the postillion unharnessed the horses and left Vignoles and his companion (John Coulthard) there all night; and six oxen were needed next morning to pull the carriage out.

When he reached home he was disturbed to find that there was no news of money from Russia; in fact the Ambassador had received instructions to stop all payments until further notice; and from Kiev it was reported that money released in St Petersburg some weeks previously had not yet arrived. He now therefore decided to return to Russia at the end of Jan-

uary 1851, hoping to be able to use his influence with the Tsar to settle
the question of payments once and for all. One of the points at issue was
the question as to which currency the money should be paid in.

On arrival at St Petersburg Vignoles was informed that the matter of
the payments was under consideration by the Council of Ministers. While
waiting for the Council's decision he received a call from a senior officer
of Kleinmichel's department, who instructed him on how he was to press
his claim. 'It was clearly understood between us that I was to be *grateful*
for his advice.' The Council of Ministers at last decided that over £15 000
should be paid in London in sterling; but at the same time Vignoles was
made to understand that several officials of the Ponts et Chaussées department must be 'rewarded' for their 'advice', and it seemed probable
that he would only get the contracts for the other bridges by the use of
similar 'rewards'.

While waiting for the Council's decision, Vignoles was taken ill with
a serious throat infection with a high fever; and he was also involved in
an unpleasant accident when the British Vice-Consul's phaeton overturned and threw him and the other occupants out. His hip and knee were
badly bruised, but to his relief no bones were broken and he was able 'to
do without leeches'. On the other hand he managed to use the period of
his enforced stay for working on his plans for the other bridges, and preparing a draft proposal for a railway from Warsaw to St Petersburg, the
cost of which he estimated at £15 million for 700 miles. With the prospect
of future employment in Russia in mind, he had sent his son Henry away
from Kiev to spend a year studying Russian with a family in Toola, in the
centre of the country.

News from Kiev being satisfactory, he returned to London to bring
the bridge accounts up to date, and paid several visits to the Great Exhibition, where a second model of the bridge was 'very advantageously
placed, but the account of it in the catalogue very meagre'. Captain Kirchenpauer, who had been with Vignoles in St Petersburg as his adviser
and interpreter, was also there to see the wonders of the Crystal Palace,
having been invited by Vignoles to accompany him to London on a short
visit, during which he was shown the sights of London, and visited the
Thames Tunnel and Stephenson's Britannia Bridge.

At the end of June 1851, Vignoles took his wife on her first trip to Kiev,
travelling via Cologne, as he had been filling in his spare time at home
making designs for a new bridge there. At Kiev he found the works in
reasonably good order, but the water unusually high, and the staff waiting for the floods to drop several feet. A few days later Count Kleinmichel made a formal inspection; the Count was in good humour, and
spoke highly of the works at dinner afterwards, which prompted Vignoles
to make the wry comment 'this is all very well, but it draws a line up to
this present'.[3]

On 28 July a total eclipse of the Sun, visible 100 miles south of Kiev,
was to take place. Astronomy was one of Vignoles's many interests, and
he naturally seized the opportunity to observe the eclipse, taking Bourne
with him to make 'daguerrotype' photographs of the Sun during totality.

Table 4. *Kiev Suspension Bridge: principal data*

(As set out in the description of the model of the bridge displayed at the 1854 Exhibition at the Crystal Palace, Sydenham.)

Extreme length	2,562 feet
Extreme breadth	52½ feet
Main spans, from centre to centre of piers	440 feet
Side spans	225 feet
Swivel bridge opening . .	50 feet
Clear water way at highest floods	2,140 feet
Height of platform above summer water-line	30 feet
Greatest depth of water in summer	40 feet
Ditto at highest floods . .	60 feet
Extreme height of piers above foundations	112 feet
Breadth of portals piercing towers	28 feet
Height of ditto	35 feet
Chord of chains of large spans, clear of piers . . .	416 feet
Versed sine of ditto . . .	30 feet
Total length of all suspended platforms, clear of piers .	2,090 feet
Total weight of platforms and rods, clear of piers .	2,350 tons
Weight of chains and pins, suspended clear of piers .	1,076 tons
Total weight of the four chains and pins only . . .	1,578 tons
Minimum sectional area of the four chains, excluding pins and overlaps	328 sq. in.
Total weight of iron used in the works	3,500 tons
Total quantity of timber used, including temporary works	500,000 cu. ft
Total quantity of masonry, brickwork and concrete .	1,500,000 cu. ft
Proof load per available sq. ft. of platform	63 lbs
Total proof load calculated to be laid on bridge for testing	2,350 tons
Actual load laid on bridge for test, 60,000 cu. ft of wet sand, 1 cwt. per cu. ft being equivalent to weight of 50,000 infantry soldiers or about	3,000 tons

TOTAL EXPENDITURE ABOUT £432,000 STERLING

With Bourne, Mrs Vignoles, an assistant named Shaft, and a supply of telescopes and cameras, he made the three days' journey to Ooman, a town well within the Sun's shadow path. Here the party passed a restless night in an inn overrun by fleas and surrounded by howling dogs. Next day, unfortunately, the weather turned stormy, and though they had an intermittent view of the early part of the eclipse, they were smothered in sand and rain during the period of totality. Vignoles was greatly disappointed, and decided to leave Ooman that evening rather than face another night in their uncomfortable billet.

Substantial progress was now being made on the bridge, in spite of shortage of labour and the usual interruptions of saints' days; the brick-lined abutment tunnels were nearly ready to receive the ends of the chains; the portals on river piers 4 and 5 were rising steadily; the swing-bridge was in position; number 5 coffer-dam was repaired and being filled with concrete. Only dams 3 and 4 were still giving trouble, which continued all through August in spite of day and night working by relays of men to clear the layer of sand and mud from on top of the concrete in them. Once again a gale broke away part of the pontoon section of the temporary bridge, taking with it a barge carrying a pumping-engine and sinking them both in the process. Once again Vignoles wrote to Douglas Evans complaining of Schweitzer's neglect in not providing enough workmen, and warned him that the work might well continue into 1853, the costs and consequences of which he laid firmly on the contractors. However, on 9 September (a Tuesday, for as Vignoles commented 'No Russian will start important work on a Monday!') the first foundation stones were laid on the concrete in number 4 dam, and on 15 September in number 3, an event which the resident staff celebrated with champagne.

Activity rose to a height at the end of September in preparation for another official visit by the Tsar. Vignoles was especially pleased to be able to show him the various stages of the work going on simultaneously: the completed roadway on the swing-bridge; the foundation and regular granite courses being laid in dams 3, 4 and 5; the centring of the portal arch fixed on number 4 river pier (dam number 6); on number 5 pier the arch turned enough to show its architectural character (here the Tsar climbed to the top of the scaffolding to see the view, and spoke to the bricklayers); and between numbers 4 and 5 river piers and the tunnel of the left bank mooring abutment the ropes which had been hung on the downstream side to represent the position of the chains. Pleased with all he saw, the Tsar expressed his confidence that on his next visit he would walk over the finished bridge. That evening Vignoles noted his hopes that when the frosts came the masonry and brickwork of the bridge would be high enough above water for the work to be continued during the next spring floods; in which case 'I look forward with Confidence to the Completion of the Works next year'.

With unusually fine weather in October, it seemed that he might be right. The portal arches of piers 4 and 5 were both closed before the end of the month and preparations were going on for setting the saddle-stones to carry the chains. But on 31 October a disturbing discovery was made. Cracks appearing in the masonry of number 5 dam indicated that some settlement was taking place. Hutton and Kirchenpauer thought it might be due to the way in which the concrete had been put in, while dredging was still going on, and while clay puddle from between the rows of piles was leaking in, owing to the severe disturbance the piles had suffered during the 1850 floods. 'This,' Vignoles noted, 'may have caused unequal settlement in the Concrete, when the great Mass of Masonry came to be laid on, and it will be well if this be really the Cause, and not primary failing in the Dam itself, which is not impossible considering the cranky State it is in. *Nous verrons!*'

Next day the settlement was no worse, though Vignoles was worried by the possibility that the timber sides of the dam might give way altogether under the weight of masonry bearing down on them, and so bring about a total collapse. 'Kirchenpauer is very uneasy & has communicated his panic to me.' For several days the levels of all the piers were carefully checked, and Vignoles gave orders for iron ties to be put through the faulty pier at each course. On 10 November measurements revealed that an average settlement of one inch overall had taken place in the masonry of dam number 5, but on the 14th the last masonry course below the roadway level was laid with no further sign of settlement. Vignoles and his staff could breathe a sigh of relief. Now, with the coming of colder weather, the workmen were beginning to be restive, and on 16 November at the height of a heavy storm some of the masons began to leave for home, trudging away in small parties, each group with a horse and cart to carry their belongings. The bricklayers worked for a few days more until frost put an end to their work for the winter. With the first three river piers well above flood-level, the portal of number 4 completed, the saddle-stones set in number 5, and the brickwork of the tunnel abutments finished, with the ends of the bridge chains firmly anchored in them, Vignoles had every reason to be pleased.

With the work so far advanced he was determined to take every precaution against the spring floods. Before leaving Kiev for the winter he gave orders for the piers and dams to be weighted down with all available pieces of iron, and for as much rough stone as could be collected to be sunk onto the fascine mattresses. He also had a long conference with Kirchenpauer and Hutton on the method of hoisting the suspension chains. Cause for further satisfaction was a report from St Petersburg that the Tsar had declared that he had seen two wonders in the South of Russia – the Kiev Bridge and a certain Regiment of Hussars!

Vignoles was not in Kiev again until the following September (1852); but he visited Warsaw in March, where he finally gave his consent to Hutton's marriage, having been 'badgered on all sides'. Though he raised his son's salary to £400 a year, he was unable to make him any personal allowance, for his own financial position was as parlous as ever. Money was now being paid at irregular intervals from St Petersburg, but he was a long way short of recovering what he himself had advanced.

During the spring of 1852 the work entered its fifth year; the portals were completed one by one, and it was time to begin raising the chains into position over the saddle-stones. Vignoles had left the initiation of this major operation in the hands of Kirchenpauer, in whom he had complete confidence.

The task of getting these chains into position, two on each side of the bridge, the links of each being made up of eight 12-ft wrought-iron bars bolted together, each bar weighing 4½cwt, was a formidable one. We have no record of how it was done, but Vignoles refers to the extensive scaffolding required, and the usual shortage of workmen capable of erecting it. When he reached Kiev in September only one chain was in position, and it was clear that it would be impossible to complete the bridge that year.

On this occasion the photographer Roger Fenton (well-known later for his photographs of the Crimean War) accompanied Vignoles on his journey to Kiev, together with a large quantity of photographic impedimenta. John Cooke Bourne had continued taking his daguerrotypes and calotypes through the various stages of the bridge's construction. He had even made a daguerrotype of the Chief Engineer, which unfortunately has not survived. The invitation to Fenton to take additional photographs indicates the importance Vignoles attached to having a pictorial record of the work in progress, probably the first example in the history of engineering of the use of photography for this purpose.[4]

The work of raising the chains continued through the winter, whenever the weather allowed; the engineers counted on taking advantage of the ice on the river to get the chains of the centre spans hung; but although there was plenty of snow that winter there was no great frost, until after severe storms at the beginning of March. From then on the work went rapidly forward, and on 30 April 1853 the last chains were safely in position.

The end was now in sight. The Emperor had expressed a desire to open the bridge on 11 September, his heir's birthday, and the fifth anniversary of the laying of the first stone. There was much to be done, but once again, owing to a failure of the necessary advance from St Petersburg, there was hardly any money to pay the workmen. Kirchenpauer wrote regularly to Vignoles in London, declaring there was danger of a complete shut-down. However on 24 June he reported that the matter had been satisfactorily settled, and that the contractor had agreed to final completion in three months. But to make things doubly sure Vignoles raised yet another loan from Palmer, of £17 000, on the security of a life insurance and of the whole of the money due to him once the bridge was opened; and he carried a large part of this sum personally to Kiev, in gold. Mrs Vignoles went with him on this journey, during which the carriage overturned, throwing them both into a ditch.

Although shaken by the accident, Vignoles was considerably moved by the sight of the almost complete bridge, and its 'whole appearance so grand'. He found the staff making every exertion to complete the platform in time for the testing by the Bridge Commission from St Petersburg.

In an article in the *Mechanic's Magazine*, written before the bridge was completed, Vignoles referred to the platform's 'novel combinations of wood and iron' and to its extreme stiffness. He was to enlarge on the latter point in a discussion on suspension bridges at the Institution of Civil Engineers in 1867, when he explained that the two longitudinal beams carrying the cross-girders of the platform were 20-feet-deep trusses, giving the effect of 10-ft parapets above the platform, between the carriageway and the footpaths on either side, and extending to a similar depth below. The result was so stiff a construction that the heaviest gales had caused no appreciable lateral movement; and vertical movement or undulation of the platform was further prevented by fixing the suspension rods at 6-ft intervals, intermediate rather than opposite to one another.

The testing by the Commission took place on 30 September, the opening having been postponed for a month. Hundreds of sackloads of sand were brought in wagons and spread evenly over the whole platform, the weight being about 1½tons per foot of the entire length of the bridge. Vignoles was apprehensive of the result, for the Commissioners had insisted on applying a much heavier load than he had at first agreed to; in addition the weather was threatening and there was every prospect of the sand being drenched with rainwater. However the bridge stood the test well. To Vignoles's consternation the Commissioners then ordered the main load to be thrown off, leaving two thirds of the load on one span only, with no weight to counterbalance it on the other side of the piers. To make matters worse, heavy rain now fell, saturating the sand and considerably increasing its weight; whereupon the generals departed hastily, apparently alarmed at the possible consequence of their unreasonable demands. However the span and piers held firm, and Vignoles noted with relief that there was no permanent deflection of the bridge platform nor stretching of the chains. He calculated that the total load on the bridge had been equivalent to the weight of 40 000 men. The wet sand was eventually cleared two hours after midnight by an utterly exhausted team of staff and men, who were rewarded by their chief with a tot of vodka and a special bonus. On 1 October Count Kleinmichel agreed to authorise the Government to pay 150 000 silver roubles direct to Palmer in London, and Vignoles wrote reporting the success of the test to Du Plat, and to Sir John Fox, whose firm had made the chains.

To Vignoles's distress and disappointment, the Tsar was in the end too unwell to perform the opening ceremony. His place was taken by his third son, the Grand Duke Nicholas. On 10 October 1853 Vignoles wrote:

I had given all necessary directions for the arrangements last night and having invited some friends to be present, we went down from the Podol in Boats and arrived at the Bridge at 12 o'clock just as the religious ceremony by the Chief Archbishop Metropolitan of all the Russias had begun. The day was most auspicious & beautiful & while we stood on the bridge in the sun we were oppressively hot. A temporary Altar was erected on the Swivel Bridge immediately over its central pivot & some hundreds of priests of every rank and from every church and convent in Kieff were assembled around it with Choristers and all the pomp of the Greek Church. The whole of the Clergy were dressed in full state Costume & a very gorgeous sight it was, especially the procession of Monks with their Banners, pictures of Saints, Crosses & Croziers, winding down from the heights and chanting as they came. The . . . Cliff and Summit were crowded with people as well as the approach to the Bridge on the Kieff side . . . About 1 pm. the Grand Duke Nicolai arrived with the Gov. Gen. & Kleinmichel and after the Conclusion of the Church service the whole religious procession moved off, chanting, and proceeded across the Bridge to the left bank followed by the crowd so that the Bridge

Kiev Suspension Bridge from a sketch by John Cooke Bourne. A version of this sketch appears in the catalogue for the 1851 Exhibition, so it must have been made when the upper works of the bridge were barely visible. Bourne probably used the model to help him achieve this remarkably accurate forecast (cf. photograph p. 121).

was once more tested with a living load & not the smallest vibration was observable with the vast assemblage moving over the platform. The Monks halted at the Entrance of the Bridge on the left bank & the Grand Duke passed on to the Shades of the Trees on the Chaussée – his suite of carriages had followed down the Bridge & the Duke took leave of me, thanking me in the name of the Emperor, & promising to make a good report to H.I.M. and thus passed off the long expected opening, successful in every respect except in the absence of the Emperor!

Vignoles spent a few days clearing up and disposing of his furniture and other belongings, finally leaving Kiev on 16 October. When in 1854 the outbreak of the Crimean War put an end for two years to relations between England and Russia, his claims were still unpaid. Nicholas I died in 1855. At the end of 1856, with the war over, Vignoles returned to St Petersburg and was warmly welcomed by old friends; but he spent a weary and infuriating two months waiting for the Bridge Commissioners to consider the claims of himself and his partners, while every possible obstacle was put in his way. Eventually, over a period of several years, he received only about half what he was owed, and nothing at all for the work done on the designs and estimates for the other bridges, and his railway plans, none of which came to fruition.

Yet disappointment and annoyance must have been far outweighed by the satisfaction of this, his greatest, achievement, which set the seal on his position among the leading engineers of his time.

The total cost was about £432 000 sterling, nearly twice the original estimate. At the time of its construction the bridge was one of the largest of its kind in the world. Tests carried out in 1888 by the Russian Public Works Department showed that no substantial deterioration had taken place in the strength of the wrought-iron chains. It remained intact, despite the annual onslaught of ice and flood, until the year 1920, when it was blown up by the Polish army during the civil wars which followed the Russian Revolution. When it came to be rebuilt a plan was put forward for an exact restoration of Vignoles's design; but the headroom above the river and the narrowness of the channel through the swing-bridge were judged to be inadequate for modern navigational needs. As the Russian newspaper the *Red Transport Worker* put it in 1923, referring to the 'ornament and pride not only of Kiev but also of all bridge-building', it was wise for economic considerations to reject the rebuilding of the 'beautiful historic monument' in favour of a less beautiful but technically more satisfactory design. A larger and higher bridge was built, though still on the foundations which had been laid with so much difficulty under Vignoles's direction. In its turn this bridge was destroyed in the Second World War.

12   The last works, 1847–1863

The several journeys Vignoles made between Russia and England during
the years 1847 to 1853 were not only for the purpose of obtaining supplies
and money for the Kiev Bridge, but also to see to the various projects
at home in which he continued to be involved. Inevitably the number of
these had decreased. Indeed, Olinthus Vignoles, in his biography of his
father, suggests that ever since his visit to Württemberg Vignoles had
been growing more and more out of touch with engineering developments in England, and that he was distressed to find that schemes originally proposed by him had been taken over and developed by other engineers. This will perhaps explain why in his later years he undertook
more work abroad than at home. Nevertheless, the pages of his diary testify to the fact that whenever he came home from Russia he still had
plenty to do.

The North Western Railway (the 'Little North Western'), mentioned
in an earlier chapter, is a case in point. A large interest in this concern
had been acquired by the Leeds & Bradford Company, and it was agreed
that their engineer, Robert Stephenson, should act jointly with Vignoles.
However the two engineers seem to have come to some amicable agreement on their joint responsibility, according to a note in Vignoles's diary

The south-eastern
end of the
viaduct at Bell
Busk on the
North Western
Railway, which
crosses the upper
waters of the
River Aire, here
a quiet trout-
stream.

on 11 April 1845, and there is no further mention of Stephenson in connection with this line. As we have seen, Vignoles was present at the cutting of the first sod of the N.W.R. just before he left for Russia in January 1847, and he continued to keep in close touch with its affairs during his early years at Kiev. That he managed to do so was greatly to the credit of his loyal and efficient Acting Engineer, John Watson. However, it was Vignoles who drafted the specifications for the locomotives ordered from Fairbairn's, and he personally supervised the designs of bridges and stations. On 5 June 1847 he laid the first stone of the viaduct over the River Aire at Bell Busk. His three youngest sons were there, their father was presented with a silver trowel and the day finished with a cold collation and rural games.

On the Lancaster branch of this line, which diverged from the Ingleton line at Clapham, were three examples of laminated timber bridges designed by Vignoles. Two of these crossed the River Wenning, near High Bentham; one was composed of three 50-ft spans, the other of two, the spans consisting of laminated timber arches, with a rise of 9 ft 6in., the timber girders measuring 33 in. deep by 16 in. wide. The third bridge was built to carry the Lancaster to Poulton le Sands and Morecambe extension across the Lune at Lancaster. In this case the banks of the river were too low for him to adopt the arched viaduct form. His solution was to combine the girder and suspension principles in a series of laminated bows springing from clustered pile piers. The platform of the bridge was partly supported by the piers and partly by suspension rods hanging from continuous 12-in. timber rails carried by three rows of bows. The spans of the bridge averaged 60 ft, the total length being 620 ft overall. A unique feature of the bridge was that it combined the curve and the skew, crossing the river at an angle of 40 degrees in a curve of 590-ft radius. The two outside bows of the bridge were formed of layers of 10 in. by 3 in. planks, the centre one of 11 in. by 3 in. planks. The suspension rods

Vignoles's timber bridge over the River Lune at Lancaster, on the Poulton and Morecambe branch of the N.W.R. To the right, in the far distance, can be seen the bridge by which the Lancaster to Carlisle line crosses both the Lune and the N.W.R.

were of wrought iron, 1⅝in. in diameter, passing through the joists of
the bridge at intervals of 3 ft 6 in.

An article in the *Illustrated London News* for 3 November 1849, recording the opening of a portion of the N.W.R., and from which our illustration is taken, reports that the local inhabitants thought that a bridge
so light in appearance would not survive the first serious flooding of the
river. However the expectations of those gathered to see it collapse were
not realised. A more serious examination by the Board of Trade Inspec-
tor of Railways in June 1848 revealed that there was a deflection of only
¾ in. in the bridge platform under the weight of a locomotive and six wagons carrying 50 tons of rails. Vignoles's bridge survived until 1864, when
it was replaced by an iron viaduct, which in turn was replaced in 1911.

The foregoing are the only examples of Vignoles's timber bridges of
which the details survive, with the exception of the Etherow and Dinting
Vale viaducts on the Sheffield & Manchester Railway, originally designed by him, but modified and carried out by Joseph Locke. (See p. 91
above.) It will be remembered that Vignoles had favoured a flatter arch
than those used by John Green and Locke. It is interesting to note that
in the case of the High Bentham bridges the rise of 9 ft 6 in. in a 60-ft
span, though sharper than Vignoles planned for the Etherow viaduct (25
ft in a span of 180 ft), was still considerably flatter than that adopted by
Locke (40 ft in a span of 150 ft in the case of the Etherow viaduct, and
25 ft in a span of 125 ft at Dinting Vale). In a discussion of a paper presented to the I.C.E. by John Green's son Benjamin on 9 March 1841,
Vignoles declared his preference for a flatter arch (as he had already
shown in his designs for the S.&M.R. bridges), but without entering into
the reasons for his opinion. He did in fact promise to present a paper on
the subject to the Institution, illustrated by a model, but it never
materialised.

Nor have we any record of how he proposed to build the 400-ft span
timber bridge over the Suir at Granagh, on the Waterford & Limerick
Railway (see p.113 above). In this connection he consulted Brunel, who
wrote that he saw no difficulty in building spans of 250 ft or even larger,
though he did not comment on Vignoles's design. In the case of the other
large bridge he projected, over the Menai Straits (see p.101 above), we
have the record in his diary on 10 August and 15 August 1843 of two separate plans: a diagonal crossing on four arches of 250-ft span, 120 ft above
high water, and one of seven arches of 250-ft span, the cost of which he
estimated at £105 000 (£60 000 for the timber and £45 000 for the
masonry).

Vignoles records that he made yet another sketch for the crossing of
the Menai Straits in April 1845, at the request of Captain Beaufort, on
behalf of the Chester & Holyhead Railway, just before this railway acquired its Parliamentary Act and appointed Robert Stephenson as Engineer-in-Chief. The plan was for a bridge with two main timber arches
of 400-ft span, sited at the Britannia Rock, where Stephenson was ultimately to build his tubular bridge.

In a paper contributed to the British Association at its meeting at Glasgow in 1840, entitled 'On Timber Bridges of a large size, in special reference to Railways', Vignoles advocated the use of timber in the interests of economy, particularly in hilly country, where it was necessary to cross steep ravines or valleys at a high level in order to maintain lines at a steady gradient. In such conditions, where the expense of constructing stone viaducts could be prohibitive, timber might well provide the solution. He instanced Wales, Ireland and the west of England as examples. In the last area Brunel was to prove Vignoles's point in his masterly series of timber viaducts in Devon and Cornwall. These were of a horizontal rather than arched shape, generally composed of horizontal continuous laminated timber beams supported on struts fanning out from the top of masonry or timber piers. Brunel's only example of a laminated timber arch bridge was the two-span skew bridge carrying the G.W.R. over the Avon at Bath. It was typical of Brunel that he used 6-in. laminations, twice as thick as those used by Green, Vignoles and other engineers, thereby assuring an unusually stiff structure, and an unusual life of 38 years for his bridge.[1]

To return to the North Western Railway, this line, like so many railways of its time, ran into difficulties late in 1847 owing to shortage of money. Vignoles encouraged the Directors to continue the work on at least part of the line, while practising strict economy, and did his best to protect the interests of William Coulthard and George Thornton, the contractors. But as he became more deeply involved at Kiev he could spare less attention for the N.W.R., and the Company, possibly following their engineer's advice to economise, dispensed with his services without ceremony at the end of 1848. Watson was retained as Chief Engineer, and the line was completed as far as Ingleton in 1851.

While Vignoles was at work on the Kiev Bridge, his friend and rival, Robert Stephenson, was building his tubular bridges at Conway and across the Menai Straits. In 1847 part of Stephenson's girder bridge over the Dee at Chester had fallen into the river, while a train was passing over it, with the loss of five lives; and Vignoles, with Locke, Gooch and other engineers, rallied to Stephenson's support at the enquiry into the disaster. He wrote congratulating Stephenson on the opening of the Conway Bridge, and managed to join the many engineers present in June 1849, when the first tube of the Britannia Bridge was floated along the Straits before being hoisted into position; but, perhaps owing to his absence in Russia, he was not one of the group in John Lucas's painting of a conference of engineers at the bridge.[2]

The summer of 1849 was notable for a visit paid by the Queen and Prince Albert to Dublin. Vignoles seized the opportunity afforded by this visit to advertise an enterprise in which he had been interested for some months, and from which he was hoping to make some money. This was a scheme for the production of coke from bog turf. His attention had been drawn to the subject by an old friend in Berlin called Elliott,[3] who was hoping to establish a works for making coke from turf dug from the moorland country near that city. Vignoles had made Elliott a considerable advance in capital, and together with him had recently taken out a patent for the process. In principle this consisted in drying out and carbonising the turf by exposing it to steam at a high temperature. Vignoles was hoping to interest Pim and other Irish friends, for the plan seemed to him to be suitable for development in Ireland, with its scarcity of coal and enormous reserves of peat bog.

The occasion of the royal visit gave Vignoles the chance to arouse interest in his plan by fuelling the royal train on the Dublin & Kingstown Railway with coke made from Irish turf. He postponed his departure for Russia in order to deal with this; but unfortunately no Irish turf coke could be produced in time. Nothing daunted, he managed to procure twelve casks of the required substance from a turf coke works at Hamburg. This arrived at Kingstown on 3 August, two days before the royal visitors were due to arrive. Next day the fuel was tried out successfully in the locomotive *Burgoyne*, driven by Wilfred Haughton, the D.&K.R. Superintendent of Locomotives. There is no record of its being used in the royal locomotive *Cyclops*, on which Vignoles rode beside Haughton, as the Queen and Prince Albert and their four children were carried from Kingstown to Dublin; in any case the origin of the coke had somewhat spoiled his plans; but no doubt he was able to forget his disappointment in attending Her Majesty's *levée* at Dublin Castle and a review of 10 000 men in Phoenix Park. He certainly noted with satisfaction that during the week of the visit 200 000 passengers travelled on the railway without mishap, and that £3500 was taken in fares.

The subsequent history of the turf coke project was to prove disastrous. It evinced no great enthusiasm from Pim or other Irish business men, and in spite of the large sums Vignoles had advanced to Elliott, in January 1850 the latter's Berlin plant was needing still more capital, and its prospects of eventually going into production and of doing business seemed slight. Vignoles continued to canvass the idea in Ireland, reading a paper on the subject to the Royal Irish Academy, and publishing at his own expense a booklet in which he described his patent process, and produced estimates of cost to demonstrate the cheapness of turf coke in comparison with coke made from imported coal. But in April 1850 he was forced to admit that 'the Turf Affair is a failure'. A month later he withdrew from the scheme. It is possible that Elliott was too old a man to pursue the project with any vigour. It is certain that Vignoles had not much time to spare for it.

By this time Elliott was owing over £8000, just when Vignoles himself was trying to raise more than that sum to meet demands for the Kiev Bridge. In February 1851 Elliott signed an agreement to discharge his debt in four annual instalments; but only a small amount was ever paid, and the following year he was declared bankrupt. For some time he had been in poor health, and in April 1852 he died. Vignoles recovered barely enough from the Berlin lawyers to pay for the education there of his two grandsons, Charles and Walter Croudace.

Among English projects at this time, Vignoles was engaged as consulting engineer at Birkenhead Docks, and to the Liverpool & Garston Railway, and he carried out extensive surveys for a possible Central Kent

Railway, in the area served today by a network of commuter lines. He also advised Sir John Burgoyne, in the latter's capacity as Inspector-General of Fortifications, on the plans for building forts at Portsmouth against the threat of invasion by Napoleon III.

But, as we have already remarked, the main emphasis in his work was now to be on projects outside the United Kingdom. Although with the completion of the Kiev Bridge and the coming of his sixtieth birthday he might well have contemplated retirement, two factors prevented this: his lack of capital and his boundless physical and mental energy. During the next two years he was to be involved with the construction of four railways abroad. The first of these was the Frankfurt, Wiesbaden & Cologne Railway, to which he was appointed Chief Engineer in 1853. It was backed by English capital, two of the Directors being Liverpool friends of his, Rathbone and Ewart. The plan was to extend the Frankfurt–Wiesbaden line to Rüdesheim, and along the right bank of the Rhine as far as Cologne; in the end, owing to shortage of capital, the line was built only as far as Rüdesheim under Vignoles's direction, though he made the plans and sections from Rüdesheim to Koblenz. Both Hutton and Henry were resident engineers at different periods of the construction.

In 1854 Vignoles was called in as consultant engineer by the Western Railway of Switzerland, to investigate the state of the lines under construction between Geneva, Lausanne and Neuchâtel. The joint English–Swiss directorate were in dispute with their contractor, and Vignoles acted as arbitrator. He subsequently took over the design and construction of the line from Lausanne to Morges and Yverdon, with his son Henry as Resident Engineer.

At the time he was undertaking work in Spain and South America. The first portion of the railway from Bahia to the River San Francisco in tropical Brazil was built under Vignoles's direction, although he never set foot in the country; one reason for this being that for almost exactly the same period he was Engineer-in-Chief on the Tudela & Bilbao Railway in Spain.

The work of carrying out the first survey of the Bahia line devolved on Hutton, who was taken off the Wiesbaden railway and packed off to Brazil with some reluctance, since he was obliged to leave his young wife and baby daughter in England. The assignment was more than usually difficult, for it involved working without good maps, in a wild and undulating stretch of country within 12 degrees of the Equator. Hutton was obviously unhappy about his results; his father was critical as ever, but managed to fix the line of the railway on the information Hutton supplied, supplemented by a further detailed examination of the country by a geological surveyor.

The concession to build the first 76 miles of the line from Bahia was granted to an English company by the Brazilian Government, which guaranteed the interest on a capital of £1 800 000. It was to be built as a single-track line, with a gauge of 5 ft 3 in. Vignoles confirmed his essential confidence in his son by appointing him resident engineer, and John Watson became the contractor. Watson also took over the running of the first section of the line in August 1860, although the large and handsome terminus station at Bahia was not completed until a later date. A remarkably clear and informative set of photographs taken during the construction of this line survives.

All the engines, rolling-stock, rails and iron-work of these railways were supplied from England; while in the case of the Bahia railway, not only did coal have to be shipped across the Atlantic, but a large number

Photographs taken during the construction of the Bahia & San Francisco Railway, *c.* 1860.

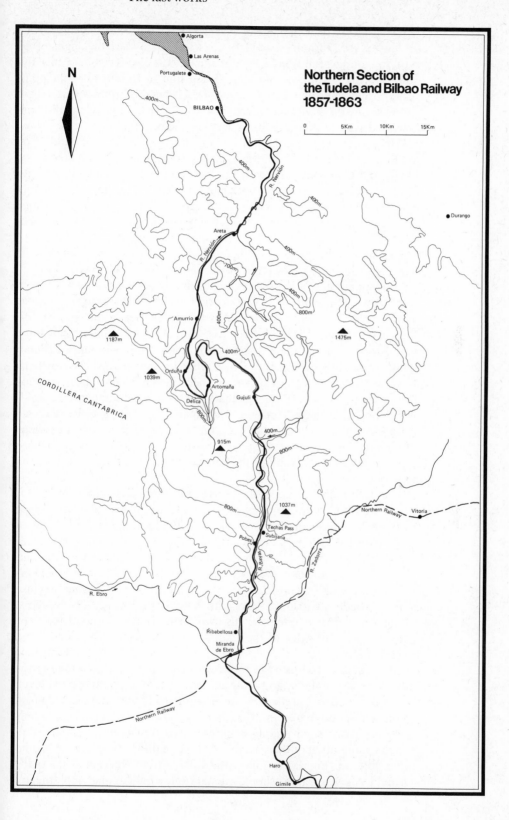

N

Northern Section of the Tudela and Bilbao Railway 1857-1863

0   5Km   10Km   15Km

Algorta

Las Arenas

Portugalete

BILBAO

400m

Durango

400m

R. Nervión

Areta

400m

R. Nervión

700m

400m

800m

Amurrio

1187m

1475m

400m

Orduña

1039m

Artomaña

Delica

Gujuli

CORDILLERA CANTABRICA

800m

915m

400m

800m

800m

1037m

Northern Railway   Vitoria

Techas Pass

Pobes   Subijana

R. Bayas

R. Zadorra

R. Ebro

Ribabellosa

Miranda de Ebro

Northern Railway

Haro

Gimile

of navvies were recruited in Northern Italy. An outbreak of yellow fever took its toll of 13 foremen and labourers in 1860, and on the engineering staff Vignoles's grandson Walter Croudace was also a victim of this disease. Hutton's wife and children joined him in Brazil, but unhappily Amélie died while he was working there. He and his father severed their connection with the Bahia railway in 1864, but in later years the line was extended as far as the San Francisco River and beyond.

While Vignoles was content to supervise the planning and building of the Bahia railway from a distance, he kept in close touch with the Directors of the Company, and with the Brazilian Government, through its Minister in London. On the other hand, in the case of the Tudela & Bilbao Railway he was fully active on the ground as well as in the drawing-office and the board-room, as the closely-written pages of his diary testify.

The first railway built on Spanish territory had been opened in November 1837 in Cuba, at that time under the Spanish Crown. In metropolitan Spain interest in railways began chiefly in the north, where there were strong links with business concerns in England. George Stephenson was there in 1845, but was put off by the difficulties of the country. By 1856 several plans for linking the northern ports with the interior were in existence, among them a survey for a line from Miranda to Bilbao made by the Spanish engineer Don Calixto de Santa Cruz. The Northern Railway from Madrid to Irun had also been begun. In 1857 the merchants of Bilbao, spurred on by the fact that their rivals in Santander were building a line to Burgos, formed a company to construct the Tudela–Miranda–Bilbao line. Capital was raised largely in the northern province of Viscaya, as well as in the rest of Spain, and even in Havana, whose merchants saw the advantage the line would bring to their trade with the mother country.

The Company commissioned a Birmingham business man, J. O. Mason, to find them an English engineer, and it was he who recommended Vignoles for the post. He and Vignoles spent a fortnight in November 1857 riding over the Cantabrian Mountains and down the valley of the Ebro as far as Tudela. On their return to Bilbao, Vignoles met the Directors of the Company and agreed to accept the post of Engineer-in-Chief, having first insisted on expunging from his contract a clause which forbade him to leave the country without the Council's permission.

The problem was one after his own heart. Barely 30 miles from the coast the summit ridge of the Cantabrian Pyrenees, rising in places to over 4000 feet, marks the watershed between the Bay of Biscay and the Mediterranean. To reach Miranda from Bilbao, this ridge has to be crossed, its lowest point being a saddle at a height of 2163 feet above sea-level, near the hamlet of Gujuli. The Spanish engineer Santa Cruz had planned to breach the ridge by a more or less direct assault, which he could only do by using very steep gradients and extensive tunnelling. Vignoles's thorough examination of the terrain led him to a different and more ingenious solution. This was to follow the valley of the Nervión as far as the Orduña 'conch' or 'horseshoe', a remarkable semi-circular amphitheatre topped by limestone cliffs 1000 feet high, which close in the head of the valley at its southern end, and from which the Nervión cascades in a fall of 700 feet. Passing within a quarter of a mile of Orduña, Vignoles's line makes a loop round the head of the valley below the cliffs, crossing the river by a viaduct, and travels back along the opposite side of the horseshoe, climbing steadily until it is separated by only 600 yards horizontally from where it enters it, though nearly 500 feet higher up. It then doubles back round the shoulder of the mountain, and winds across the heads of several steep valleys on the mountainside, to bridge another lofty water-

Viaduct at Delica, Tudela & Bilbao Railway, 1972. The head of the Orduña loop, looking north-east.

Summit viaduct at Gujuli, Tudela & Bilbao Railway, 1972. The 400-ft waterfall beyond the bridge was almost dry when the photograph was taken.

fall near the Gujuli summit. From this point, after passing through a short tunnel, there is an easy descent from the Bayas valley to the plains of Miranda and the River Ebro. On this last section some heavy excavation, including another tunnel, was necessary to carry the line through the deep gorge of the Techas Pass.

By using this route Vignoles managed to achieve a maximum gradient of 1 in 70 over the 12 miles between Artomaña and the Gujuli summit, while elsewhere the gradient does not exceed 1 in 100, and the average is about 1 in 200. Such gradients in the mountainous section of the line were only obtained with the help of sharp constantly reversing curves, a measure for which Vignoles was criticised by the Spanish government engineers who were required by law to approve the plans, but which he persuaded them was quite acceptable. As he pointed out, the costs likely to arise from extra wear and tear on rails and rolling-stock would be considerably less than those caused by steeper gradients or more extensive cuttings or embankments.

After crossing the Ebro at Miranda the line follows the right bank of the river the whole way to Tudela. For this reason the gradients are inconsiderable, but the elaborate windings of the river, and the steep cliffs below which it flows as it curves its way through the rocky countryside, necessitated heavy earthworks in places, accompanied by similarly sharp curves.

For the initial survey, Vignoles assembled a staff of 13 assistants and adopted his usual method of deploying them at different points along the line, while he set up a drawing-office in Bilbao from where he could direct the whole operation. The work of supervision took him many times over the ground; although now into his sixties, he had lost none of his capacity to spend long days in the saddle or jolting in his carriage over rough roads. He quickly established good relations with the Chairman of the

The Techas Pass, Tudela & Bilbao Railway, looking south, 1972. The road tunnel has been built since the railway. Traces of the old road can be seen at the foot of the cliff above the river.

Table 5. Tudela & Bilbao Railway. Abstract of the costs of construction

Localities	Lengths		Cost of land and works			Average cost per			Observations
	English miles	Metrical kilom.m.	Spanish (reals vellon)	French (francs)	English (£ sterling)	Spanish league (reals vellon)	French kilometre (francs)	English mile (£ sterling)	
Bilbao to Arrancudiaga	10	16,044	25 Millions	6,578,950	263,158	8,726,000	410,056	26,316	Exit from Bilbao
Artómaña to Yzarra	14	22,512	37 Millions	9,736,850	389,474	9,253,730	432,518	27,820	Heaviest part through Pyrenees
Amurrio to Artómaña Yzarra to Poves	22	35,388	25 Millions	6,578,950	263,158	3,956,140	185,909	11,952	Lightest part through Pyrenees
Amurrio to Poves	36	57,900	62 Millions	16,315,800	652,632	5,996,550	281,793	18,129	Passage through Pyrenees
Arrancudiaga to Amurrio Poves to Miranda	$19^3/_5$	31,556	13 Millions	3,421,040	136,842	2,307,000	108,412	6,982	Lightest works on 2^a section
Bilbao to Miranda	$65^3/_5$	105,500	100 Millions	26,315,790	1,052,632	5,308,060	249,438	16,046	First section (up to the Ebro)
Miranda to Alcanadre	$60^6/_6$	97,845	60 Millions	15,789,500	631,580	3,434,000	161,372	10,388	Passage along the rougher country
Alcanadre to Tudela Junction	$28^3/_{10}$	45,468	15 Millions	3,947,375	157,895	1,847,450	86,817	5,580	Passage along the plains
Miranda to Tudela Junction	$89^1/_8$	143,313	75 Millions	19,736,875	789,475	2,930,680	137,719	8,858	Second section (down the Ebro)
Bilbao to Tudela Junction	$154^{39}/_{40}$	248,813	175 Millions	46,052,632	1,842,105	3,938,700	185,089	11,886	Whole line

The above table is taken from the pamphlet describing the relief model of the Cantabrian Pyrenees, and of the route followed by the Tudela & Bilbao Railway over the mountains, which was shown at the International Exhibition of 1862 (see pp. 162–3).

The table has two errors. In the column headed 'Lengths' the total for the whole line should read 154²⁹/₄₀ miles. In the column headed 'Observations', 'First section' should read 'Second section' and vice versa.

Company, Don Pablo de Epalza, a leading Bilbao business man, and with the Managing Director, Don Segundo Montesina, both of whom had spent many years in England. Epalza was Chairman of the Bank of Bilbao, which gave considerable backing to the railway.

The contract to build the northern section of the line, as far as Miranda, was awarded in September 1858 to Thomas Brassey, who was at the time in partnership with Sir Joseph Paxton, the architect of the Crystal Palace. Brassey was one of the leading English railway contractors. He and Vignoles had a great respect for each other's talents, though they did not always see eye to eye over the details of the work. Brassey claimed that Vignoles had underestimated the difficulties of the country, particularly as regards the proportion of rock to earth which had to be shifted, while Vignoles criticised Brassey's resident agent, Bartlett, for lack of energy and drive. One point where Vignoles had to give way was over the number of viaducts to be built on the section between Orduña and the summit. Henry Vignoles, who became Chief Resident Engineer, wrote of this section:

On this part of the line for a distance of about 30 kilometres, the works were the heaviest. Deep cuttings, high embankments, sharp curves, tunnels, viaducts and colossal retaining walls following each other in constant succession. . . . Most of these [the viaducts] had to be built across the deep lateral valleys high up on the mountainside where there was not a semblance of a road.[4]

The problem was to get the right stone up to the sites of the viaducts, for the limestone of which these mountains were composed was not suitable for the purpose. After he and Brassey had clambered about for a couple of days among snowdrifts, Vignoles reluctantly agreed to sacrifice a number of his 'works of art' in favour of high embankments, themselves major feats of engineering, pierced by massive culverts. The reduction in time and expense afforded by these changes was all the more necessary, in view of the continual interruptions to the work from two causes unforeseen by Brassey – the first the heavy rainfall hardly expected in Spain, the second (not unfamiliar to Vignoles after his experience in Russia) the large number of saints' days observed as holidays by the workmen. Arthur Helps, Brassey's biographer, calculated that because of these two factors only 200 days in the year could be worked.

Helps gives an interesting account of the problems Brassey had to contend with in paying his men. As Spanish business men were unused to cheques, the cash for the monthly pay-out of 10 000 men had to be accumulated until it weighed over a ton. It was then sent out from Bilbao in a large coach drawn by six mules, escorted by a couple of civil guards, the clerk being armed with a revolver. Surprisingly, no attempt was ever made to rob the coach, though on one occasion an axle broke under its weight. There was, however, an alarming incident at Orduña, when Brassey's agent in the town was besieged for a whole night in his house by one of the sub-contractors, who happened to be a Carlist chief; the latter had got into difficulties over paying his men, and threatened to kill everyone in the house unless he was paid a sum much greater than was due to him. He was only persuaded to raise the siege when a detachment of soldiers arrived, summoned from Bilbao by one of the railway staff who had managed to escape from the town on horseback.

Brassey completed his contract within the stipulated time, though at a heavy financial loss. Henry Vignoles wrote 'dear old Tom, instead of sending in a large bill for contingencies, merely remarked that it was not always possible to make large profits'; however, when it came to the contract for the Ebro section of the line, 'dear old Tom' set his terms so much higher than Vignoles's estimates, that neither he nor the Company were prepared to agree to them. Vignoles tried to interest John Watson (just home from Bahia), and other English contractors, but in the end the work was undertaken by a Spanish firm, and eventually completed under the direction of the Company engineers, who had been joined by Philip Sewell, one of the engineers on the Santander line.

On the Ebro section, a stretch of the river near Logroño ran for some way at the base of perpendicular cliffs several hundred feet high. To avoid having to bridge the river, or to take the line out of the valley, which would have involved expensive cuttings and steep gradients, Vignoles decided to blast away the cliffs for about half a mile, and to divert the river. Remembering his experience at Salthill on the D.&K.R., he enlisted the advice of Sir John Burgoyne in choosing the position of shafts and galleries in the cliffs, and on the amount of gunpowder to be laid in them. For the diversion of the river he drew equally on previous experience; for he accomplished it without driving a single pile, by using fascine mattresses filled with earth and stone to form a dam, having specially engaged a large team of Dutch workmen for the operation. Both these projects were at first strongly opposed by the government engineers, and a local newspaper declared that the works would never be completed, but they remain as a memento of English temerity.

Vignoles was never more in command of his powers or more thoroughly in control of the whole situation than during the building of this railway. The country appealed to him, with the sharp contrast between its well-watered slopes and lofty limestone precipices on the northern side of the mountains, and the more barren and torrid outcrops to the south. He was still undaunted by the prospect of sleeping on the floor, or at best on a 'tolerable' truckle-bed, in a Spanish country inn, while out on a survey in the mountains. His grasp of detail, as illustrated in the pages of notes and calculations in his diary, remained unimpaired. On the other hand there are now more frequent references in the diary than in former years to fatigue and short bouts of illness. Much of this was undoubtedly due to long overland journeys to and from London. One such journey is worth recounting in detail. On 1 February 1860, he left Bilbao by coach at 9 p.m., travelled all night and the following day through the mountains to Bayonne, in hail, sleet, thunder and rain, meeting Montesina and the government engineer for a business discussion on the way. He left Bayonne at 7.30 next morning, by train for Bordeaux, where he changed stations and had two hours to wait. He continued his journey by the Paris mail train, dined at Angoulême on the way, and reached Paris at 5.30

a.m. on the 4th. He called for breakfast and letters at the Hotel Bristol, his usual halting-place, and so missed the train for Calais by one minute, after which he spent the day on small items of business in Paris, until 'Having met a Biscayan Captain of a Merchantman, without a cloak this cold weather, I lent him mine and a Rug, and got him safely off by the 5 p.m. Express for Cologne and Hamburg.' He himself left by the 7.40 p.m. train and had a good crossing, arriving at Dover at 4.30 a.m., where as it was a Sunday he had to wait three hours for a slow train which landed him at London Bridge at 11.45. The rest of the day he worked in his Duke Street office at an accumulated back-log of papers about the Bahia railway and other business, besides the plans he had brought with him to discuss with Brassey and his partners.[5] Small wonder that even he should confess to feelings of fatigue.

The journey to Bilbao was of course easier by sea, though it could take longer. Vignoles used this route when he took with him his wife and Hutton's two children, Isabella and Charles, who had been sent back from South America after the death of their mother in Bahia. Elizabeth maintained a home for her husband and the children in or near Bilbao for nearly four years and, besides playing her part as a hostess, went with him by carriage on several of his trips into the mountains. Social life, embracing both local society and the engineering staff, followed a pattern which had been established in Kiev, a pattern which included Church of England services each Sunday and daily family prayers for the whole household. On New Year's Day 1860, we read in the diary:

This morning, in consequence of the absence of the Clergyman, the Rev. Mr. Chapman . . . I read the prayers of the Church of England, in the temporary room fitted up by Mr. Bartlett at the Contractor's Offices – and afterwards read part of a Sermon on 'Labour' from the published Volumes of the Rev. Mr. Bellew. In the afternoon, this being New Year's Day, we had a number of the Assistants to a friendly Family Dinner.

The benevolence of the Engineer-in-Chief towards his staff took other forms. The disapproval of Vignoles of Brassey's chief agent, Bartlett, to which we have already referred, came to a head in a dispute between him and Henry Vignoles about a division of responsibilities, in which Bartlett was patently in the wrong. He refused to admit it until, on Vignoles's complaining to Brassey, he was ordered to apologise, which he did with a somewhat ill grace. It happened that Vignoles had promised his assistants a holiday 'to play at Cricket', and he conceived the happy idea that the contractor's Staff should also take a holiday and end the dispute with a friendly match against the engineers.

The match was to take place a few days later, on a field near the river-mouth, and Vignoles, after writing to all the district engineers to come in for the game, promised to arrange for the company steam-launch to take the players to the ground. Unfortunately an event of far greater importance supervened. This was the arrival off the mouth of the Nervión, on 9 July 1860, of H.M.S. *Himalaya*, a troopship bearing a party of astronomers and scientists who had come to view the total eclipse of the sun on 18 July. The expedition had been organised by the Royal Astronomical Society, with government support.

It must have been a cause of great satisfaction to Vignoles that the shadow-path of the eclipse passed directly over his railway. He was a friend of the Astronomer Royal, Professor G. B. Airy, and was himself a Fellow of the Royal Astronomical Society. As soon as he learned of the proposed expedition he had offered to place himself and his staff at their disposal and set about preparing a map showing the shadow-path, and a short guide-book to the country to go with it, to which Professor Airy added an appendix of technical instructions. On 8 July 'after having almost broken my heart and worn out everybody to get my arrangements complete', Vignoles received a telegram saying the *Himalaya* was arriving a day earlier than expected. Thus it was that instead of being available for the cricket match that day, the company steamer and other boats went off to ferry the passengers ashore from the ship lying outside the bar, together with their boxes of instruments, telescopes and other impedimenta. Vignoles had arranged for various members of the staff to act as hosts, and his own house was stretched to the limit. Among its ten guests were Olinthus Vignoles and his wife Mary. Olinthus was now a fully-fledged parson, and during his stay was much in demand to baptise a number of babies born to the wives of members of the engineering and contractor's staff, among them his brother Henry's first-born son.

The astronomers were advised by Vignoles to place their observation posts on the southern side of the Cantabrian range, away from the sea-mists which form so often on the northern slopes. Among those who took his advice was Warren de la Rue, who had come to photograph the Sun's corona and prominences, and whose 30 packages of photographic apparatus and portable observatory, weighing in all about two tons, were conveyed by Bartlett's staff over the mountains to Rivabellosa, near Miranda. Two days before the eclipse there were violent thunderstorms and strong winds, and the morning of the 18th was misty and overcast. However, at Rivabellosa the clouds cleared, and de la Rue and his team of observers obtained the first historic photographs of the Sun's prominences. Professor Airy at Gujuli also had an excellent view, but Vignoles was once more unlucky, for the 5000 feet height of the Gorbea mountain to which he had climbed remained wrapped in cloud; and it was to his 'great mortification' that he found, on returning to Bilbao, that the eclipse had been plainly visible there.

The astronomers and their ladies spent a further week sightseeing, and sorting out and comparing their results. For days Vignoles had hardly been able to give a thought to the railway except when showing it off to his guests. Fifteen or more people sat down to dinner in the little house at Albia each day, and the company passed the evenings with conversation and music, in the pleasant aroma of their host's excellent Havana cigars. By 26 July, when the *Himalaya* was due to leave for England, Elizabeth and her staff were worn out with their exertions, and Vignoles, who was returning with the party to England and trying to settle his affairs before leaving, 'finally gave it up as the hurry and excitement in my last duties to the Astr. Party were too much *even for me*'.

In his 'Bakerian Lecture' to the Royal Society on the results of his observations, de la Rue paid tribute to the generous hospitality shown to the expedition by Vignoles and his staff. Vignoles records that de la Rue took many photographs of both hosts and guests in Bilbao and on the voyage home. Unfortunately it has not been possible to trace any of them.

While the railway was still building, Vignoles followed his usual practice of having a model made, which in this case was a panorama in relief of the mountainous section of the line, from Orduña to Miranda. This was exhibited at the International Exhibition of 1862, and subsequently

The Techas Pass, Tudela & Bilbao Railway, looking north. From a drawing by Percival Skelton, published in the *Illustrated London News*, 1863.

presented to the Royal Museum in Madrid. In the pamphlet describing the model Vignoles pointed out, for the interest of students of military history, that the railway passed through the Techas Pass by the route followed, during the battle of Vitoria, by a strong detachment of Wellington's troops in a movement to outflank the French. The map also showed the position of the house near Subijana where the Duke lodged for the night before the battle.

As a soldier, and one who in his professional life had had several contacts with Wellington, Vignoles was naturally deeply interested in this campaign; but nowhere in the pamphlet, nor indeed anywhere in his diary and correspondence for this period, is there a word to suggest that he himself had been there, or that he had ever set foot in Spain before 1857.

The railway was opened from Bilbao to Miranda in March 1863, the ceremony including an episcopal blessing of the locomotives.[6] The remaining section was completed in the following year. Vignoles had made plans for an extension as far as Portugalete and had surveyed the estuary with a view to building a harbour in the mouth of the river; but the Company were at the time too hard-pressed for funds to undertake further ventures, though his plan was ultimately carried out by Spanish engineers some time later.

Vignoles's line survives to this day virtually unchanged.[7] The northern section was built to take a double track, but for most of its length the single line as laid down by Vignoles remains. From Bilbao to Miranda the line has been electrified, and a frequent service of railcars runs every day between these two cities, taking the gradients and curves over the Gujuli summit and round the Orduña horseshoe with an ease which Vignoles would have approved. Less frequently, heavy goods trains drawn by grimy black steam locomotives toil high along the mountainside, and once or twice a day the line carries the coast to coast passenger train from Barcelona to Bilbao. To an English observer its grey stone viaducts remain its most striking feature, for they seem to have been lifted straight from some Lancashire valley, and contrast strangely with the Spanish village churches, or the red-roofed cottages glimpsed between their elegant piers.

13 The end of a career, 1863–1875

With the completion of the Tudela & Bilbao Railway Vignoles's active engineering career was virtually at an end, although he was involved both financially and as a consultant in work on which his sons Hutton and Henry, now senior engineers in their own right, were engaged: Henry as Thomas Brassey's partner in the building of the Warsaw & Terespol Railway, and later as Chief Engineer to the Isle of Man Railway; Hutton in Leipzig, where he was Chief Engineer of the Leipzig Tramways Company.

But to the end of his life his interest in engineering and scientific subjects continued unabated, as evidenced by his frequent participation in debates and discussions at meetings of the various scientific societies to which he belonged. In 1855 he had been elected a Fellow of the Royal Society. His grandfather, also a Fellow, was distantly related to Sir Isaac Newton, and in 1841 Vignoles had presented to the Society the portrait of Newton by Vanderbank which had belonged to Dr Hutton. With Roger Fenton he was one of the founder members of the Photographic Society of London, now the Royal Photographic Society. At the first meeting, held in 1853, Vignoles remarked on the value of the new art to engineers, referring to the photographs taken during the construction of the Kiev Bridge, especially those which showed the method of raising the chains. It is tantalising to read in his diary of an exhibition of these photographs which he organised in London in January 1855, for except for two prints by Roger Fenton in the Royal Photographic Society's collection, no trace has been found in England of any photographs taken during the building of the Kiev Bridge. As Vignoles mentions having discussions with Bourne at this time, it seems likely that the exhibition consisted mainly of prints of Bourne's calotypes. The Prince Consort was a visitor, and was sufficiently interested to request copies of one or two of the photographs. (Vignoles took the opportunity of letting the Prince know 'that the Emperor of Russia had not paid for the Bridge although he had been using it for 15 months!') In the same month Vignoles was making enquiries about making collodion glass slides from paper negatives for projection with a lantern, and travelled to Paris to see an exhibition of stereoscopic photographs projected by electric light.

His interest in astronomy was lifelong. In January 1829 he was elected a Fellow of the Royal Astronomical Society, of which he was Vice-President from 1863 to 1865. In December 1870 he demonstrated his continuing interest by making a third attempt to witness a total eclipse of the Sun. He and Hutton joined the R.A.S. expedition to Sicily, Vignoles himself having been responsible for all the arrangements for the overland journey via Munich and the Brenner Pass to Naples. Unfortunately H.M.S. *Psyche*, in which the party embarked for the crossing to Sicily, struck a hidden rock off the coast of the island, and became a total loss, although there were no casualties. Olinthus Vignoles relates that his father, who was working in his cabin at the time, could only be persuaded to leave when the water was covering the floor, quietly remarking that it was more than half a century since his *first* shipwreck on the island of Anticosti.

C. B. Vignoles as President of the Institution of Civil Engineers, 1870. Oil-painting by his daughter, Camilla Croudace.

Misfortune dogged the expedition to the end, as bad weather broke out, and for the third time Vignoles was unlucky.

At the end of 1869 he attained the crowning honour of his profession, when he was elected to the Presidency of the Institution of Civil Engineers. Although he had served for many years on the council and as a Vice-President, he had missed his turn to be elected to the chair, largely because of his frequent absences abroad. His election in his seventy-seventh year was a fitting recognition by his fellow engineers of his position as a senior member of his profession.

His presidential address, delivered on 11 January 1870, is very much the testament of an elder statesman, ranging widely over the field of engineering from the earliest times, and in many different countries, and looking back over his own personal experiences to the beginning of the railway era. He had in fact so much to say that parts of the address, as afterwards printed, had to be left out. Much of the printed version consists of a series of comments, supported by statistics, on such diverse subjects as the history of the French ministry of Ponts et Chaussées, the English canal system (with a sidelong glance at Indian roads), the Suez Canal, the Bermuda floating dock, armour plating and rifled ordnance, Blackfriars Bridge, Prussian dockyards and the Mont-Cenis Tunnel. No doubt an audience accustomed to sitting through the enormously diffuse speeches which graced Victorian public occasions took it all in its stride. We might have found it more difficult. To us, the most interesting passages, apart from those which deal with Vignoles's own experiences, are those concerning the differences between the French and English system of training engineers, with the French emphasis on theory and the English on practice; and the section in which he recalls the rejection of the recommendations of the Irish Railway Commission, and regrets the uneconomic and piecemeal development which resulted from it. Although an individualist who resented any kind of official interference in what he was undertaking, Vignoles was (as we have seen) always ready to see the advantages to be gained by centralised planning. And he illustrates his point by again drawing a comparison between the English and French systems of free enterprise and centralisation.

This was the only paper of any size that Vignoles contributed to the proceedings of the Institution, though he intervened frequently in the discussions, giving his opinions on a wide variety of subjects. Thus, his experience at the Woodhead Tunnel prompted him to join in a discussion in 1864 on the sinking of artesian wells, when he pointed out that a steam-driven boring machine manufactured by Mather & Platt of Manchester was in principle similar to the tool he had used on the Sheffield & Manchester Railway works, itself based on a method used by the Chinese 1500 years before. In 1865 he reported his observations on the use of concrete in Mediterranean dockyards, which material he had been informed was liable to deteriorate, possibly because of some particular property in the water of that sea. In 1866 he contributed a historical note on the origin of the word 'camel' applied to a contrivance used for raising sunken vessels, which arose from the use by a Russian sea-captain, in the time of

Peter the Great, of an old hulk called *Camel* as a floating dock. Also in 1866, in a discussion on the water supply of Paris, he commented on the fact that half the supply was absorbed in fountains and other ornamental displays, while the poorer people living at the top of tall buildings had to go without; he therefore considered that the supply of water to Paris was a failure, though successful as an engineering operation. In the same year he recommended the installation of high platforms in railway stations, citing the great risk and fatigue suffered by passengers as a result of the very low platforms at many continental stations.

His interest in the engineering problems connected with the movement of water prompted him to intervene in the discussions on the building of sea defences and banks to protect reclaimed land; on the influence of breakwaters and piers in the prevention of scouring; and on the force and action of wave motion; and he entered into discussions on the purification of water supplies, the drainage of London and Paris, and the treatment of town sewage. Finally, in February 1870, in a discussion on railway income and expenditure, he put forward the unusual view that landlords ought to give their land free to railway companies, since he maintained that the value of their estates was considerably increased by having railways constructed through them.

These are but a few examples of Vignoles's many contributions to the proceedings of the I.C.E. We have already referred to the lectures he gave in 1840 and 1842 to the British Association (see Chapter 8 above). In 1857, after the completion of the Kiev Bridge, he read a further paper to the Association 'On the Adaptation of Suspension Bridges to sustain the passage of Railway Trains'. In this he concluded that it was quite feasible to use suspension bridges for carrying railways, provided that due consideration was given to the rigidity of the platform and the prevention of undulation, vibration and oscillation. In this connection he referred to the steps he had taken at Kiev to ensure rigidity by the depth of the platform, and to counteract oscillation and undulation by the alternating disposal of the suspension rods (see p. 143 above).

As we have seen, Vignoles was frequently called upon as a Parliamentary witness. His professional advice was also sought by other official bodies. For instance, in 1848 he was called as a witness at the sessions of the Gauge Commission, appointed to resolve the difficulties arising from the disparity between the 7 ft gauge of Brunel's Great Western Railway, and the 4 ft 8½ in. used in the rest of the country. In his evidence Vignoles stated that in the various lines for which he had been responsible in England he had adopted the 4 ft 8½ in. gauge, mainly because they were adjacent to other lines of this gauge. He himself would have preferred a gauge of 6 ft. He considered this would provide greater stability in running, and make possible a more efficient design of locomotive, and larger carriage body. On the Eastern Counties line he had intended to use the 6 ft gauge, but after discussion with Braithwaite a 5 ft gauge had been adopted. A few continental lines were adopting a broader gauge, but the majority were following the example of the United Kingdom, since their locomotives and rolling-stock came from there. He had remended 4 ft 8½ in. for Württemberg, and for the above reason the Directors of the Dublin and Kingstown Railway had chosen the standard gauge, instead of the 6 ft gauge he originally recommended. But while he preferred the larger gauge he did not think its advantages were sufficient to warrant a change now that the 4 ft 8½ in. was so widely established.

Although the Commission voted in principle in favour of the narrow gauge, the battle between the two systems continued, and they existed side by side until after the death of both Vignoles and Brunel. Where traffic had to pass from one system to another the transfer was at first effected by completely unloading and reloading passengers and baggage, a process which caused considerable delay and confusion. Later the problem was solved to a certain extent by adopting a 'mixed gauge', i.e. laying a third rail between the rails of the broad gauge lines, for the passage of narrow gauge trains on them. Both Brunel and Vignoles made suggestions which would have eased the problems of transfer. Brunel's scheme anticipated the modern use of box containers, which could be transferred from one train to another; Vignoles proposed that narrow gauge wagons should be run up onto broad gauge wagons, following the early example of gentlemen's carriages, or French 'diligences' which were carried, passengers and all, on the Paris–Rouen railway.

In the field of draftsmanship and mapping Vignoles had few equals, and it was fitting that in 1855 he should be appointed to the Royal Commission on the Ordnance Survey. He had often been handicapped in his work by inadequate maps, and was one of the first engineers to indicate heights on his own; and he strongly recommended the general use of contours by the Survey.

Apart from official consultations, he was always ready to offer an opinion on problems which took his fancy, such as the question of the proposed improvements to navigation at the mouth of the Danube in 1856, on which he prepared a lengthy paper for Lord Palmerston and Sir John Burgoyne, to whom the conclusions of the International Commission on the subject had been referred. It is also more than probable that he was the author of an anonymous pamphlet outlining proposals for building new government offices in Whitehall, and a new Westminster Bridge.

In 1867 Vignoles acquired a pleasant country house at Hythe, on the shores of Southampton Water, where he and his wife settled down to a life of semi-active retirement. The Villa Amalthea had its own flagstaff, a telescope on the terrace, and a boat slung in davits from the garden wall.[1] Here Vignoles, now a Justice of the Peace, led the life of a country gentleman, visited frequently by his family and friends. Since his eldest son's mental illness, the duties of the heir had devolved on Hutton, who had married again in 1860 while in Bahia, his 18-year-old wife being Carolina Schleusner, daughter of a German ship-broker and Portuguese mother in Brazil. As in all Victorian families, new Vignoles children arrived on the scene with great regularity, and there were always babies and young children about the house and garden to cheer the hearts of the old man and his faithful companion. Two or three times a week he would tra-

vel up to London to his office in Duke St , Westminster, where he also
rented a suite of chambers as a town residence.

Throughout his life Vignoles had enjoyed very good health, though his
early adventures had left him subject to colds and recurrent bouts of
fever, from which he invariably seemed to recover remarkably quickly.
He had more than his fair share of nearly fatal accidents, and in his old
age suffered a good deal from lumbago. But in spite of a life as full of
hard unremitting toil and anxiety as theirs, he outlived his three great rivals, Brunel, Stephenson and Locke, by more than 20 years, thanks to

C. B. Vignoles in
old age, *c.* 1874.

a quite unusually robust constitution.

The end when it came was mercifully quick. In the middle of November 1875, Vignoles spent three nights in London 'on the spree', as he told Hutton and Henry, with whom he played whist one evening in his chambers, while on the other two he dined with the Engineer Volunteer Corps (in which he held the rank of Lt. Colonel), and the Royal Astronomical Society. On the evening of his return to Hythe he suffered a severe stroke, and died four days later, on 17 November 1875.

He was buried on 23 November in Brompton Cemetery, attended by three sons, three grandsons, and many of his friends and colleagues in science and engineering, both military and civil.

What position does Vignoles occupy in the history of engineering? For all his undoubted ability, his actual achievement, in terms of engineering work completed, seems modest compared with that of Brunel, Locke or Stephenson. No great trunk line fit to compare with the G.W.R. or the L.&N.W.R. stands to his credit. Those lines he did complete were not big enough to survive on their own, and were soon absorbed into larger systems. Yet it fell to his lot to lay the foundations of two most important projects, each of which was completed by another man: the Liverpool & Manchester Railway by George Stephenson, the Sheffield & Manchester by Joseph Locke; and having built the first railway in Ireland, he was only prevented by political decisions beyond his control from becoming Engineer-in-Chief to the southern section of the great trunk system envisaged by the Irish Railway Commission. None the less the great bridge at Kiev and the Tudela & Bilbao Railway were works which could stand comparison with any contemporary achievement in England.

In the opinion of O. S. Nock, by comparison with the 'Great Triumvirate' Charles Vignoles was a most unlucky engineer. And he certainly had an unfair share of misfortune. The disaster that overtook him as Engineer-in-Chief of the Sheffield & Manchester Railway seemed at that moment to wreck a career which had offered unlimited promise. The arrangement which Vignoles entered into with the Directors of the S.&M.R. was certainly a peculiar one; but however rash he may have been in investing so heavily in the company's fortunes, he can hardly be blamed for expecting the Directors to keep their word. The event illustrated three outstanding traits in his character: his unbounded enthusiasm, his liberal and carefree attitude towards money, and the naïve innocence which led him to expect from others an integrity equal to his own.

His enthusiasm impelled him to throw himself heart and soul into any project he undertook, and to allow nothing to turn him aside from its completion; and he was understandably impatient with those who stood in his way, or showed signs of wavering in the face of difficulties. With enthusiasm went an unusual quality of vision, which caused him invariably to see a project as part of a much bigger whole. Such were his dreams of a railway system in Ireland, or a trunk route from the Channel to the Mediterranean. As for money, he regarded it as a means to an end. Although he set his estimate of the value of his service high, no man was ever less mean or parsimonious. Generous to a fault, he was as ready to lend as to borrow, and however severe his criticism of his family or of those who worked for him might be, they could always be sure of his support in time of need.

In his Presidential Address to the I.C.E., Vignoles quoted Napoleon's dictum 'Engineers ought to have magnificent ideas', a saying which aptly fits his own character. He might have added that they should also be magnificent people. Vignoles's houses at Dinting Vale and Kiev, his lavish hospitality, his articles in the Press, his expensive models, these are not just the expressions of an extravagant nature with a liking for 'cutting a dash', but symbolise his idea of the engineer's rightful position in the social hierarchy. He was himself a gentleman in every sense of the word, and a man of honour; and he seems to have had a natural social grace which enabled him to move with ease in Court, diplomatic and government circles. It is perhaps characteristic that in the pages of his diary the worst that he can say of a man is that he is guilty of ungentlemanly behaviour. As a gentleman he was of course by no means free from the prejudices of his class; but though suspicious of revolutionaries and political reformers, he believed firmly in the need for improvement of human conditions, and he had the capacity to be moved by the misfortunes of others.

Vignoles was endowed with an unusually quick, keen and imaginative mind. This was both a help and a hindrance to him, for while it meant that he worked quickly, and could take a broad view of a problem, it led him sometimes to overlook details or jump to conclusions without sufficient examination of the issues involved. On the other hand he was rarely satisfied with the first draft of a plan or report, and would often revise it four or five times. Quick in mind, he was also quick in temper, and while finding it hard to move at a more sensible man's pace, he found it harder to suffer fools gladly. This undoubtedly made him enemies. Physically a small man (at the time of the Rainhill trials he weighed only nine stone), like all small men of strong personality he must often have exhausted his colleagues by his sheer ebullience. Yet this same ebullience, and the instinctive reaction of the small man to the buffets of a large and hostile world, underlay the resilience which brought him through disaster to final success. And he had many friends too; one of the strongest points on the credit side being the long and faithful service given to him by his assistants. He may have bitten their heads off at times, but they stayed with him, and must therefore have liked working for him.

In the course of research no first-hand candid comment on Vignoles's character has come to light, apart from the somewhat fulsome compliments paid him on his election as President of the I.C.E., and after his death, although there is no doubt much truth in the references to his geniality, courtesy and liberal hospitality in his mature age. Fortunately Vignoles himself had the capacity for self-knowledge and self-revelation to a remarkable degree; and the simplicity, honesty and idealism which were the essential qualities of his character shine through the pages of his letters and diary.

'Si monumentum requiris, circumspice.' How much of Vignoles's work remains for those who look around? In England, certain stretches of the Midland Region which have escaped the Beeching axe; in Ireland, the Dublin & Kingstown and some other lines; parts of the Trent Bridge remain; and the Ribble Bridge, though obscured by modern additions; in Russia the great Kiev Bridge has vanished, though Vignoles's plans and Bourne's water-colours and calotypes are still preserved in Leningrad, together with the Tsar's model, dismantled and packed away; the Tudela & Bilbao Railway remains the best complete example of his achievement, in a countryside virtually unchanged since the line was built. In the I.C.E. library, and in the British Transport Historical Records and other libraries up and down the country, a few plans and papers survive.

'Vignoles Bridge', Coventry, in its original position.

'Vignoles Bridge' in its new site in Coventry.

Among these it is interesting to note that it was Vignoles's report of May 1837 on the state of the S.&M.R. works that was handed to the Chairman of the British Transport Commission, by a representative of the contractors, at the opening of the third Woodhead Tunnel in June 1954.

In the United Kingdom, as Nock has pointed out, the most abiding monument to his work is the Vignoles rail, widely used on the Continent, and recently introduced in a modified form by British Rail. Yet few, even among railwaymen, now remember the name: and with an ironical twist that Vignoles himself would have savoured, on its introduction someone wrote to the Press protesting against the use of a rail invented by a *foreign* engineer.

Yet one more relic has recently come to light. On the quiet reaches of the Oxford Canal, near Coventry, a number of small cast-iron bridges, of simple but attractive design, still carry the tow-path across the loops cut out of the canal by Vignoles in his survey of 1828. One of these was endangered by the building of a motorway; but an enlightened City Engineer decided that it was of sufficient artistic and historical merit for it to be lifted bodily from the site in 1969 and re-erected across a stream on a housing estate near the centre of Coventry. It is reasonable to suppose that the bridges, which were all of one pattern and cast by the Horseley Coal & Iron Company, were designed by Vignoles; and it is as 'Vignoles Bridge' that this one is recorded in the files of Coventry's City Engineer. It is ironic that such a delicate structure should have outlived the great suspension bridge at Kiev. But such are the fortunes of war.

NOTES

Chapter 1

1. The line goes back to Captain Vignoles's great-great-grandfather, Major François la Balme de Vignoles, who died in the siege of Alicante, on 3 March 1709, during the War of the Spanish Succession.
2. Victor Hugues, Commissioner General to the National Convention, for the Windward Islands.
3. *Gentleman's Magazine*, October 1794.
4. John Bruce: Memoir of Charles Hutton, and Dictionary of National Biography.
5. British Library. Additional MSS 27899.
6. With one exception. In a letter to Olinthus Vignoles, dated 27 November 1883, J. O. Mason, who accompanied Charles Vignoles on his first visit to the site of the Tudela & Bilbao Railway (see p. 154), wrote that he recollected Vignoles saying he had not been in Spain since the siege of San Sebastián. This began on 25 July 1813 and ended on 31 August 1813, when Vignoles must have already been at Sandhurst.

    In another letter, dated 11 August 1814, Mrs Hutton urges Vignoles 'to think over the transactions of the past fourteen months'. This would put the quarrel in June 1813. The battle of Vitoria was on 21 June 1813.
7. An interesting account of the action, by Lt. Dunbar Moody, is quoted by Leask & McCance.

Chapter 2

1. In the Map Room, British Library, and in the County Library, Charleston, South Carolina.
2. St Augustine City Minutes, 20 July 1821 (St Augustine Historical Society).
3. A copy is in the British Library.
4. British Library Map Room. B.M. 72455(3).

Chapter 3

1. Son of John Hanson, Lord Byron's solicitor.
2. Vignoles to Riddle, 14 January 1827. R. E. Carlson states that 'Vignoles believed he was co-engineeer with Stephenson.' Vignoles never made this claim.
3. See also Simmons: 'A Holograph Letter from George Stephenson', *Journal of Transport History*, September 1971.

Chapter 4

1. The account of this correspondence is based on copies of letters written by Vignoles to Brunel and to Riddle. The originals have not been traced.
2. The other two machines entered for the trials were Brandreth's *Cycloped* and Burstall's *Perseverance*. The first was propelled by a horse trotting on a kind of

treadmill, and was not considered as a serious entrant. The wagon carrying the second machine to the trials overturned, and it only reached the course on the last day of the contest, and was withdrawn by its owner after a short demonstration run.

3.   This was the Whiston Plane, east of Rainhill. It was raining heavily and the five locomotives were unable to tackle the gradient until the load was reduced by ordering the male passengers to walk.

4.   Vignoles's specification reads: 'The Wheels to be 54 inches diameter, Axles 5 inches diameter, Cylinders 13 inches, Stroke 20 inches, weight with water in the boiler not to exceed 8 tons, the Engine to work with coal and the 4 wheels to be moved. The Boiler guaranteed to generate steam to a power of not less than 2500–lbs at a velocity of 10 miles per hour giving a capability to the Engine of Drawing 150–tons including its own weight and that of the tender at a rate of 5 miles per hour up an inclination of 1 in 400.' (From J. S. Allen's unpublished work; he comments: 'A very big engine for the time.')

Chapter 5

1.   I am indebted to Mr John Burnett for indicating the evidence which points to Cole as architect of the Cloncurry 'pavilions'.

2.   Mr K. A. Murray, of the Irish Railway Record Society, an authority on the history of the D.&K.R., made valuable comments on the original draft of this chapter, in one of which he pointed out that owing to its lack of funds the D.&K.R. Company could only be sure of obtaining a loan from the Board of Public Works after a drastic revision of the estimates.

3.   Olinthus Vignoles quotes these letters in full in his biography of his father, as well as a letter from Vignoles to Thiers. I have not been able to trace the originals.

4.   See Chapter 6, pp. 75ff, for a full discussion on the method of laying rails and Vignoles's views on the subject.

5.   The use of the word 'mention' is probably an understatement!

6.   It seems that the Princess had declined the invitation to be present. Charles Forth, writing to Vignoles from Enniscorthy on 3 September 1834, refers to a rumour that 'the Cholera has frightened the Princess Victoria from her intended visit to Ireland'.

7.   A copy of this map is in the British Library.

Chapter 6

1.   A pamphlet extracted from Vignoles's reports is in the Barnstaple Public Library. The canal was never built.

2.   For the subsequent history of the *Novelty* see the *Engineer*, 26 January 1906.

3.   O. J. Vignoles records the consternation of the North Union Chairman, Sir Hardman Earle, when Vignoles presented him with the proposed plan of Preston Station, at an estimated cost of £6000. The Chairman protested that it was three times too large, to which Vignoles replied that they would live to see it ten times too small.

4.   The line has been doubled and there are now two bridges and two tunnels.

5.   See also Kirby & Laurson, who state that Stevens had the rails rolled at the Guest foundry at Dowlais, Wales, and that Vignoles may have 'gotten' the idea from Stevens. They also state that Vignoles's rail was broader and flatter than Stevens's.

6.   In a discussion at the I.C.E. on 11 November 1862 Vignoles cited as an example the re-laying of the Taunus Railway in Germany, where many rails had broken a short distance from the point of support, after being turned over. He had seen 'hundreds of tons' of these rails stacked away beside the line.

7.   This statement seems to imply that the Vignoles rail had been in general use in France for some time. This is not the case. French engineers, like the British, at first tried various types, including flat-bottomed rails, and different kinds of chair. The Vignoles rail was first *exclusively* used by the Chemin de Fer du Nord, and the Chemin de Fer de l'Est, from about 1865, with the development of steel rails. In due course other networks followed suit.

Chapter 7

1.   The Sheffield & Manchester Railway Directors soon regretted their decision to build only a single-line tunnel. The second tunnel, driven on the north side of the first, was begun in 1847, and took five years to build. After nearly a hundred years of use, the vast increase in traffic and the difficulties of maintaining the two tunnels necessitated the building of the new double-line tunnel, which was completed in 1954, at a cost of £4 250 000.

Chapter 8

1.   For the full story of the atmospheric railway and the consideration of whether it could have been developed into a viable system the reader is referred to the excellent accounts by L. T. C. Rolt in his biography of I. K. Brunel, and by Charles Hadfield in his *Atmospheric Railways*, to both of whom the author is indebted for much of the information in this chapter.

2.   John Rennie, in a letter to Vignoles dated 24 October 1842, quoted a report from his Principal Engineer: 'On Tuesday the 4th day of October the Satellite took the ¼ past 4 train down from Croydon to Brighton in *54* minutes including 3 stoppages of *3 minutes each*, i.e. performed with 6 carriages the 40½ miles in 43 minutes. I am not anxious that this fact should be known to too many or they might accuse us of over-driving, but I assure you we went as steadily as if we had been going only 20 miles an hour. I hope to take the Queen down with her next month.'
     Vignoles's visitor on this occasion was M. Teisserenc de Bort, of the French Ponts et Chaussées department, for whom an extensive tour was organised, including a visit to the atmospheric railway, on which he reported to the French Public Works' Minister.

3.   In 1942 the snuff-box was presented by Mr Ernest B. Vignoles and Lt. Col. Walter A. Vignoles, grandsons of C. B. Vignoles, to the Royal Society, 6 Carlton House Terrace, where it is circulated at meetings of the council.
     Vignoles's report to the King of Württemberg, dated 30 December 1843, is to be found, together with some of his rivals' observations, in a pamphlet 'Die erste Section der Württembergischen Eisenbahnen', pub. Stuttgart, Verlag der J. B. Metzlerischen Buchhandlung, 1844. British Library, catalogue no. 8235, l.42 (1–8).

Chapter 9

1.   Lady Resident of Queen's College, Harley Street.
2.   On 17 March 1846, Vignoles and Brunel drove to Canterbury, rising at 4 a.m. next day to check borings for a tunnel at Boughton, after which they returned to London, arriving in time for Vignoles to be cross-examined from 12 until 4 – 'by which time I felt pretty well fatigued, having suffered with the cold in the early part of the day'.
3.   Vignoles had first become interested in India in 1842, when he produced a report on a proposed Central India railway, at the instigation of a barrister named P. R. Welsh. He completed the report in September of that year, after six weeks' work, gathering information from East India House, the Geological Society and the Asiatic Society. A copy was printed as Appendix 2 to the report of the I.C.E.

discussion (in which Vignoles took part) of a paper by W. T. Thornton on 'State Railways of India', on 4 February 1873 (*I.C.E. Proceedings* **35**, 475–519).

4.   Charles Forth was one of Vignoles's earliest assistants – possibly the first. His name first appears as 'Charles' in the diary entries concerning the survey for Cheltenham Water Works in March 1825 (see Chapter 3 above). Forth's wife died shortly after her husband, and Vignoles was much concerned in setting up and maintaining a fund for the assistance of their children.

Chapter 10

1.   In his unpublished autobiography Olinthus Vignoles mentions letters written to him by his brothers 'telling how nearly their noses and toes were frost-bitten, and how their stock of sherry in the pockets of my father's travelling carriage became masses of colourless ice with a scanty remnant of the original alcohol at the bottom of the bottle'.

2.   A note added to the diary by Olinthus Vignoles describes this as 'An astounding & unjust Contract!!!'

3.   Vignoles continued to be deeply concerned by his son's condition. His daughter-in-law, Martha, was reconciled with the family, and Vignoles wrote very kindly to her. Her husband outlived his sister and brothers, dying in 1908 at Henley-in-Arden.

4.   R. S. Palmer was senior partner in the firm of Palmer, Eland & Nettleship, subsequently Eland, Nettleship & Butt.

Chapter 11

1.   From O. J. Vignoles's unpublished autobiography, which has interesting passages on this visit. He was met by Hutton in Warsaw, and on the journey to Kiev the post-horses bolted and their carriage was overturned, throwing the occupants out. Luckily the two brothers escaped with a few bruises.

2.   *Pozzolana* was the name given to hydraulic cement made at Pozzuoli, near Naples. Its main constituent was based on volcanic ash.

3.   i.e. Kleinmichel wished Vignoles to forget all the difficulties he had suffered to date.

4.   In a report delivered to the 25th Congress of the International Scientific Film Association, in October 1971, the late Sir Arthur Elton stated that he saw a set of 29 of these photographs, in a faded condition, together with the Bourne sketches, at the Obrazov Institute of Railway Engineering, Leningrad.

Chapter 12

1.   The author is much indebted to Professor L. G. Booth's careful study of laminated timber bridges (see Bibliography).

2.   Vignoles crossed the Britannia Bridge for the first time on his way to Ireland on 1 April 1850, when he wrote that he was 'greatly gratified at its complete success, which *I* had never doubted. In fact I was the first engineer who erected a bridge on this principle (the hollow beam) over the Blackburn, Darwen & Bolton Railway near Blackburn over the Leeds & Blackburn Canal and the adjacent Turnpike Road abt. 4 yrs. ago.'

3.   Vignoles first met Elliott in London in February 1835, when he discussed the subject of railways in Hanover with him and the Hanoverian Ambassador.

4.   Henry Vignoles's account is in the British Library MS Department (Add. MSS 34536 f. 42).

5.   Duke Street, Westminster, now demolished to make way for Whitehall government buildings. A more fashionable site than Trafalgar Square, and also favoured by I. K. Brunel, and the young Disraeli.

6.   Beyer, Peacock & Co. supplied eight 4-4-0 outside cylinder tank engines and a number of 2-4-0 tender engines in 1862. One of the first locomotives is preserved at Bilbao.

7.   This paragraph is based on a visit made by the author in 1972.

Chapter 13

1.   Vignoles celebrated the 'house-warming' by taking his three sons and some friends on a hired steam-launch to lunch on board the 'turret-ship' *Wyvern*, as guests of Captain John Burgoyne, R.N., son of Sir John Burgoyne. The Fleet was assembled in Spithead for the Naval Review in honour of the Sultan of Turkey and the Khedive of Egypt, which took place next day in a severe gale. Captain Burgoyne perished in 1870 in the foundering of the ill-fated H.M.S. *Captain*.

BIBLIOGRAPHY

The following are the principal works consulted while preparing this book. The author is indebted to the writers named.

Ahrons, E. L. *The British Steam Railway Locomotive, 1825–1925*. London, Locomotive Publishing, 1927.

Banco de Bilbao. *Un Siglo en la Vida del Banco de Bilbao*. Bilbao, 1957.

Barman, Christian. *Early British Railways*. London, Penguin Books, 1950.

Barnes, E. G. *The Rise of the Midland Railway*. London, Allen & Unwin, 1966.

Baughan, P. E. *North of Leeds*. Roundhouse Books, 1965.

Beamish, Richard. *Memoir of the Life of Sir Marc Isambard Brunel*. London, Longman, Green, Longman & Roberts, 1862.

Briggs, Asa. *The Age of Improvement*. London, Longmans, 1959.

Bruce, John. *Memoir of Charles Hutton*. Newcastle, Hodgson, 1823.

Bryant, Arthur. *The Age of Elegance*. London, Collins, 1975.

Carlson, Robert E. *The Liverpool & Manchester Railway Project, 1821–1823*. Newton Abbot, David & Charles, 1969.

Carter, E. A. *A Historical Geography of the Railways of the British Isles*. London, Cassell, 1959.

Chappell, M. *British Engineers*. London, Collins, 1942.

Church, W. C. *Life of John Ericsson*. London, Sampson Low, 1890.

Clements, Paul. *Marc Isambard Brunel*. London, Longmans, 1970.

Close, Colonel Charles. *The Early Years of the Ordnance Survey*. Newton Abbot, David & Charles, 1969.

Coleman, Terry. *The Railway Navvies*. London, Hutchinson, 1965.

Couche, C. H. *Permanent Way, Rolling Stock & Technical working of Railways* (translated by J. N. Shoolbred and J. E. Wilson). Paris, 1877.

Dendy Marshall, C. F. *A Centenary History of the Liverpool & Manchester Railway*. London, Locomotive Publishing Co., 1930.

– *A History of the Southern Railway*. London, Southern Railway Co., 1936.

Die erste Section der Württembergischen Eisenbahnen. Stuttgart, Verlag der J. B. Metzlerischen Buchhandlung, 1844. British Library, catalogue no. 8235, 1.42 (1–8).

Dow, G. *The First Railway between Manchester & Sheffield*. London, L.N.E.R. Publications, 1945.

– *Great Central* (Vol. 1). London, Ian Allan, 1959.

Drewry, C. S. *A Memoir of Suspension Bridges*. London, Longmans, 1832.

Duff, David. *Edward of Kent*. London, Stanley Paul, 1938.

Ellis, C. Hamilton. *British Railway History* (Vol. 1). London, Allen & Unwin, 1954.

Elton, Arthur. *British Railways*. London, Collins, 1945.

Fortescue, Sir J. F. *History of the British Army*. London, Macmillan, 12 vols., 1910–30.

Hadfield, Charles. *Atmospheric Railways*. Newton Abbot, David & Charles, 1967.

Hardy, P. Dixon. *Thirteen Views on the Dublin & Kingstown Railway*. Dublin Penny Journal Office, 1834.

Helps, Sir Arthur. *Life and Labours of Mr Brassey*. London, Bell & Daldy, 1872; repub. Bath, Adams & Dart, 1969.

Hoetzch, Otto. *The Evolution of Russia*. London, Thames & Hudson, 1966.

Kirby, R. S. & Laurson, P. G. *Early Years of Modern Civil Engineering*. Yale University Press, 1932.

Klingender, F. D. (ed. Elton). *Art & the Industrial Revolution*. London, Evelyn, Adams & Mackay, 1968.

Lambert, Richard S. *The Railway King*. London, Allen & Unwin, 1934.

Lampe, David. *The Tunnel*. London, Harrap, 1963.

Leask, J. C. & McCance, H. M. *Regimental Records of the Royal Scots*. Dublin, Alex. Thom. Ltd., 1915.

Levinge, Sir R. *Historical Records of the 43rd Light Infantry*. London, 1868.

Marshall, John. *The Lancashire & Yorkshire Railway*. Newton Abbot, David & Charles, 1969.

Milligan, John. *The Resilient Pioneers*. London, Elastic Rail Spike Co., 1975.

Mills, R. *Atlas of South Carolina*. Philadelphia, A. Finley, 1825.

Moyaux, Auguste. *Les Chemins de Fer autrefois et aujourd'hui*. Brussels, Dupriez, 1905.

Murray, John. *Handbook for Travellers in Spain*. London, John Murray, 1888.

Nock, O. S. *The Lancashire & Yorkshire Railway*. London, Ian Allan, 1969.

– *The London & North Western Railway*. London, Ian Allan, 1969.

– *The Railway Engineers*. London, Batsford, 1955.

Nye, R. B. & Morpurge, J. E. *History of the United States*. London, Penguin Books, 1970.

Paine, Larran. *Bolívar the Liberator*. London, Robert Hale, 1970.

Rennie, Sir John. *Autobiography*. London, H. & F. N. Spon, 1875.

Report of the Gauge Commission. H.M.S.O., 1846.

Reports of the Irish Railway Commission. H.M.S.O., 1837, 1838.

Rolt, L. T. C. *Isambard Kingdom Brunel*. London, Longmans, 1957.

– *George & Robert Stephenson*. London, Longmans, 1960.

– *Thomas Telford*. London, Longmans, 1959.

Sidney, Samuel. *Gauge Evidence*. London, Edmonds & Vacher, 1846.

Smiles, Samuel. *The Life of George Stephenson*. London, John Murray, 1857.

Tebbutt, R. *A Guide or Companion to the Midland Counties Railway*. Leicester, 1840.

– *The Midland Counties Railway Companion*. Nottingham, R. Allen; Leicester, E. Allen; 1840.

Vignoles, C. B. *Observations upon the Floridas*. New York, Bliss & White, 1823.

Vignoles, O. J. *Life of Charles Blacker Vignoles*. London, Longmans, Green, 1889.

Warren, J. G. H. *A Century of Locomotive Building by Robert Stephenson & Coy 1823–1923*. Newcastle, Andrew Reid, 1923.

Weaver, Lawrence. *Story of the Royal Scots*. London, Country Life, 1915.

Webster, N. W. *The Grand Junction Railway*. Bath, Adams & Dart, 1972.

– *Joseph Locke, Railway Revolutionary*. London, Allen & Unwin, 1971.

Williams, E. T. & Palmer, H. M. (eds.) *Dictionary of National Biography*. Oxford University Press, 1971.

Wrottesley, Hon. G. *Life & Correspondence of F.M. Sir J. F. Burgoyne, Bt*. London, R. Bentley & Son, 1873.

Unpublished Material

Allen, J. S. *Early Railway Locomotive Building at the Horseley Coal & Iron Company, Tipton, Staffs*. 1961. Copy available in Stafford Record Office.

Booth, Dr L. G. *Laminated Timber Arch Railway Bridges in England and Scotland, 1835–1855*. (A paper read to the Newcomen Society, 13 October 1971).

Brunel, Marc Isambard. *Diaries and Reports during building of Thames Tunnel*. I.C.E. Library.

Dublin & Kingstown Railway, Minutes & Letter-Books. Coras Iompair Eireann, Dublin.

Irish Railway Commission Letter-Books. Public Records Office, Dublin.

Osborne, Richard Boyse. *Diary*. National Library, Dublin, MSS 7888–95; 9606.

Railway & Canal Companies' Minute-Books, Reports, etc. British Transport Historical Records Dept., Public Records Office, London.

Vignoles, C. B. *Diaries, 1825, 1826, 1830–62*. MS Dept., British Library.

– *Correspondence*. Portsmouth Record Office.

Vignoles, K. H. *The Infant Ensign*. MS Dept., British Library, 1967.

Vignoles, O. J. *Autobiography*. In possession of Vignoles family.

INDEX